Adhesives and Coatings Technology

ADHESIVES AND COATINGS TECHNOLOGY

Premamoy Ghosh
Former Professor
Department of Polymer Science and Technology
Calcutta University
Kolkata

Tata McGraw-Hill Publishing Company Limited
NEW DELHI

McGraw-Hill Offices
New Delhi New York St Louis San Francisco Auckland Bogotá Caracas
Kuala Lumpur Lisbon London Madrid Mexico City Milan Montreal
San Juan Santiago Singapore Sydney Tokyo Toronto

Tata McGraw-Hill

Published by Tata McGraw-Hill Publishing Company Limited,
7 West Patel Nagar, New Delhi 110 008.

ISBN (13): 978-0-07-0077195
ISBN (10): 0-07-0077193

Managing Director: *Ajay Shukla*
Head—Professional and Healthcare: *Roystan La'Porte*
Publishing Manager: *R. Chandra Sekhar*
Jr. Sponsoring Editor: *Nimisha Goswami*
Manager—Sales & Marketing: *S. Girish*
Product Manager: *A. Rehman Khan*
Controller—Production: *Rajender P. Ghansela*
Asst. General Manager—Production: *B. L. Dogra*

Typeset at The Composers, 260, C.A. Apt., Paschim Vihar, New Delhi 110 063 and printed at Gopsons Papers Ltd., A-2&3, Sector 64, Noida 201 301

Cover Design: Kapil Gupta, Delhi

RZXCRRDXRBRZA

The McGraw-Hill Companies

To

The young band of students in science and technology who stand to benefit immensely by reading and using this book

Preface

Adhesives and Coatings Technology combines and integrates two different, yet closely related areas of science and technology. These are (i) material surfaces, surface preparation and bonding of two like or unlike surfaces, using a film-forming adhesive interlayer and, (ii) surface coverage, protection and decoration using traditional and modern film-forming, adhering polymeric materials in the form of lacquers, varnishes, different types of solvent-based and/or aqueous paints and other surface coatings.

The book provides a comprehensive description on principles and practices of modern adhesives, coatings, microencapsulation and plating processes. It also covers areas like coated textiles, powder coating, tribology and tribological coatings. This book additionally complements my earlier two books, viz., *Polymer Science and Technology—Plastics, Rubbers, Blends and Composites*, 2nd ed., 2002, and *Fibre Science and Technology*, 2004, both published by Tata McGraw-Hill.

Inspiration for writing this book came from successive batches of my inquisitive students of the Department of Polymer Science and Technology, Calcutta University. Generations of students and professionals have been in focus while designing the contents of this book. I have endeavoured to treat and explain different topics in a lucid language as far as possible. It will be a matter of immense pleasure if the book fulfills the expectations of all concerned.

I express my deep gratitude and appreciation to my wife Suparna for her forbearance, encouragement, understanding and help during the preparation of the manuscript. I acknowledge in grace and with thanks the initiative, encouragement and support of the entire team of McGraw-Hill Education (India) Pvt. Ltd.

PREMAMOY GHOSH

Contents

Chapter 1

Basic Considerations

Two major application areas of polymers relate to spreading them over a specific substrate surface in the form of a thin layer, web or film; the polymer may be conveniently used in the form of a solution, paste, dispersion, emulsion, melt or powder. Monomers or oligomers may also be used in some cases. Such applications may be conveniently classified as:

(i) Adhesives
(ii) Coatings

In general, an adhesive is a substance applied as a thin, cohesive intermediate layer between the surfaces of two substrates, like or unlike, so as to bind them together by surface attachments combining adsorption, chemisorption, and probably mechanical interlocking too, often aided by heat and pressure.

Surface coatings are applied as thin films or layers; they are spread and stuck on a surface. Finally, they appear as thin, colourless or colourful, glossy, attractive and tack-free, dry, cohesive films, commonly offering a good coverage, which is usually protective and also sometimes decorative in nature. It is the surface that catches the eye. We need to save the surface to save an object. "If you have saved the surface you've saved them all" is the slogan driving the development of the surface coating industry and the spread of coating applications.

1.1 Surface Features and Forces

For a stable solid or liquid phase, the molecules in the surface find themselves in a different surrounding than those in the interior. A molecule in the interior is surrounded on all sides by neighbours with much the same intermolecular attractions in all directions as shown in Fig. 1.1. However, the intermolecular forces on surface molecules remain unbalanced as there are very few molecules in the gas phase above the surface. The surface molecules are thus subject to a net inward pull or attraction.

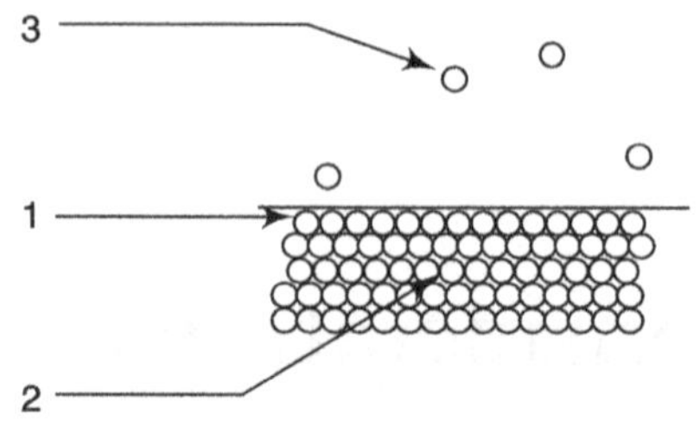

FIGURE 1.1 Schematic distribution and packing of molecules in a (solid/liquid) substrate: (1) at the surface layer, (2) in the interior and (3) in the (upper) gas phase

Therefore, there is a state of unbalance or lack of saturation in the surface. In an effort to extend the surface area, work must be done to draw molecules from the interior to the surface, against and overcoming the inwardly directed attractive forces. The surface molecules thus acquire greater energy than the interior molecules; this excess energy that they are in a position to exercise is known as the *surface energy*.

1.2 Surface Tension

Under a net inward attraction, the surface always tends to contract to the smallest possible area. Under the general trend of a system to assume an equilibrium condition of minimum free energy, the system would strive to turn to a state of minimum surface, much in tune with the external restraints prevailing. It is no wonder then that a drop of a liquid or a gas bubble appears as a sphere, a configuration that ensures minimum surface for a given volume.

For its inherent tendency to contract, a liquid surface appears to be under tension called *surface tension*. The surface tension acts in the surface plane and is expressed as the force exerted per unit length perpendicular to the direction of the force. Surface tension and free surface energy both have identical values and are useful even though the latter is more basic conceptually.

1.3 Interfacial Tension

It follows from the discussion so far that the surface of a given mass of matter behaves differently from its interior parts. The surface, as a high-energy region, is particularly consequential in cases of adsorption, wetting, dispersion, adhesion, foaming, emulsion, floatation, liquid penetration into solids, the activity of exchange resins, heterogeneous catalysis and corrosion phenomena. For two immiscible liquids in contact, the molecules of each exert an attraction on those of the other at the *interface*; this partly contributes to overcoming the inward pull of each liquid phase upon the molecules of its own kind finding a place at the interface. Thus at the interface, the molecules of each kind are under different kinds and levels of forces from those in the bulk of the two liquid systems; the surface energy at the interface, known as the *interfacial tension,* is always less than that of the phase having higher surface tension. For completely miscible liquids, there is no interface; and so the interfacial tension is zero.

Interfacial tension is usually higher between phases of dissimilar structure and polarity. Lowering interfacial tension promotes wetting, foaming, spreading and dispersion, emulsification, etc.

1.4 Surface Active Agents

Since the surface is a region of higher free energy, it is thermodynamically imperative that any component or agent that reduces the surface tension will tend to cluster or concentrate in the surface of the liquid (melt or solution) to which it is added. Such an agent, the *surface active agent,* contributes to lowering the surface or interfacial tension and enabling the breakup of an immiscible liquid phase into fine droplets in a different liquid mass of the dispersion medium, forming an emulsion often aided by mechanical agitation. Soaps or detergents are common emulsifying (surface tension lowering) agents used for the dispersion of oil in water, with which it is immiscible. Emulsions are produced in large volumes and are of much importance in high polymer practice.

The most common structure of effective, commercial surface active agents is that of a molecule possessing one or more strongly polar sites attached to the main bulk, usually a long non-polar chain unit and they may be tailored to suit specific needs. An efficient surface active agent is required to be used in very small concentrations (in parts per thousand, ten thousand or hundred thousand).

1.5 Concept of a Clean Surface

A freshly cleaned surface is highly prone to the adsorption of surrounding molecules or reacting with them. An *ideally clean surface,* which is chemically identical to the interior of the phase, is seldom available; one often needs to cope with gases and chemicals adsorbed from the atmosphere. For surface coatings and adhesives as well as plating processes, the actual condition of the surface is a vital factor. Paper and glass surfaces exhibit low electrical resistivity due to the continuous layer of adsorbed water or moisture present. The polar nature of the paper or glass surface may be changed to a non-polar one by treatment with a linear silicone polymer, viz.

$-\!\left(\mathrm{Si(CH_3)_2{-}O}\right)_n\!-$, so as to provide an array of $-\mathrm{Si(CH_3)_2{-}CH_3}$ linkages on the surface with methyl groups directed outwardly, as in Fig. 1.2.

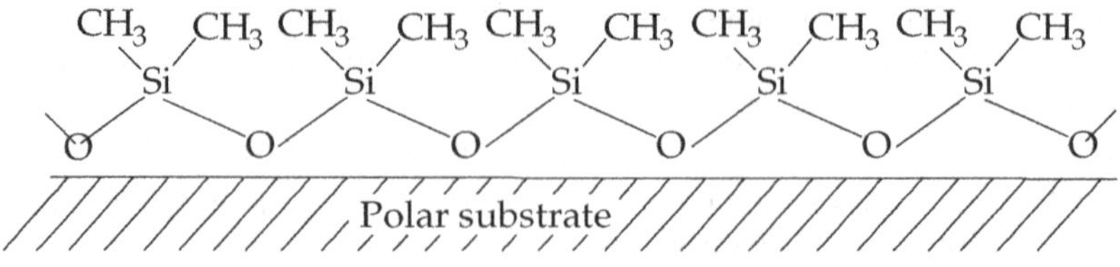

FIGURE 1.2 Array of 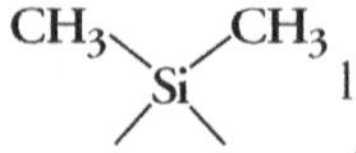 linkages on the surface of a polar substrate given a coat of silicone polymer, 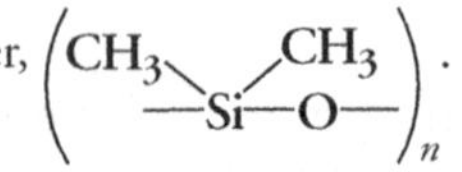.

The surface so modified is virtually hydrocarbon in character and becomes moisture repellent; hence, the electrical surface resistivity shoots up. This is a widespread technology adopted to render papers and fabrics rain-proof and water-repellent, such as for manufacturing light fabrics for umbrellas.

1.6 POLYMERS AND THEIR CHARACTERISTICS

Polymers are large molecules with molecular weights of the order of 10^3—10^6. Our fundamental needs of food, clothing and shelter are met by high polymers. Our body tissues are built up of high polymers and the life process itself involves high polymer reactions to a considerable extent. Many important materials—natural (natural gums, polysaccharides, body proteins, wool and hair, natural rubber, natural silk, animal hydes and skin, rocks and clays, gelatin, silica or sand, etc.) and synthetic (polyethylene, polypropylene, nylon polyamides, polyimides, epoxy resins, polycarbonates, rayons, alkyds and polyesters, etc.)—are high polymers. All rubbers, fibres and the bulk of plastics, surface coatings and adhesives are based on polymers or high polymers.

1.6.1 Macromolecular Concept

A polymer molecule contains many (tens, hundreds or thousands of) 'mers' or chemical units linked successively with covalent bonds with one another. The degree of polymerization (DP) is the number of mers or units (often called repeat units) in the polymer molecule. The polymer's resinous or gummy character, indefinite property range and the lack of a structural concept posed problems for their understanding till about 1920. The macromolecular concept, spelling a chain-like structure for a polymer molecule as propounded by Staudinger[1] in 1920, cleared the accumulated confusions of the earlier decades.

1.6.2 Chain Length and Molecular Weight

A large number of simple molecules from which a polymer molecule may be formed are commonly known as *monomers*. The formation of a molecule of polyethylene by the polymerization of the monomer ethylene may be simply expressed as:

$$\underset{\substack{\text{Monomer}\\ \text{(Ethylene)}}}{nCH_2{=}CH_2} \xrightarrow{\text{Polymerization}} \underset{\substack{\text{Polymer}\\ \text{(Polyethylene)}}}{—(CH_2—CH_2)_n—} \tag{1.1}$$

Here, a large number of ethylene (olefin) monomer molecules are shown to form a long, saturated polymer molecule called polyethylene (the simplest polyolefin) characterized by a chain length or degree of polymerization (DP) of n. The molecular weight of the polymer is given by the product of the weight of the repeat unit —(CH_2—CH_2)— and the degree of polymerization, n. A polymer is usually characterized by a melting (temperature T_m) range often spread over 3–10°C and is also characterized by a second order transition or glass transition temperature (T_g).

1.6.3 Molecular Weight Distribution

It has been observed that when high polymers are synthesized, all their molecules do not grow to the same size. Even though the natural polymers are relatively uniform in size, there still remains a considerable spread in the sizes and molecular weights of individual molecules. Molecular weight or size distribution is usually narrow in natural polymers and wide, narrow or very narrow for synthetic polymers.

It is imperative to remember that even a rigorously pure high polymer mass does not possess a precise molecular weight such as M_1 or M_2 or M_3 in the usual sense; actually, it is characterized[2] by an average molecular weight $\bar{M}$ such that

$$\bar{M} = f_1M_1 + f_2M_2 + f_3M_3 + \ldots + f_iM_i + \ldots = \Sigma f_iM_i \tag{1.2}$$

where, $\Sigma f_i = 1.0$

Molecular weight (size) distribution may produce significant impacts on physical and mechanical behaviours. Two different samples of a given high polymer material may exhibit the same average molecular weight and yet differ considerably in mechanical property profile and processing ease. One may consist of molecules with intermediate sizes or chain lengths and a narrow size distribution, while the other may be a blend of very low and very large molecules, thus giving a wide molecular weight distribution.

1.6.4 Basis of Average Molecular Weights

Number Average

The number average molecular weight ($\bar{M}_n$) of a given polymer sample is expressed as a summation series[2] as follows:

$$\begin{aligned}\bar{M}_n &= \frac{\Sigma N_iM_i}{\Sigma N_i} = \frac{N_1}{N}M_1 + \frac{N_2}{N}M_2 + \frac{N_3}{N}M_3 + \ldots \frac{N_i}{N}M_i + \ldots \\ &= \Sigma f_iM_i \end{aligned}\tag{1.3}$$

where N is the summation or total number of molecules present (i.e. ΣN_i) and $M_1, M_2, M_3,$ etc. represent the molecular weights of polymer molecules corresponding to $N_1, N_2, N_3, \ldots$number of molecules that constitute the polydisperse polymer. Thus, $\bar{M}_n$ is given by the summation of the products of the mole fraction of polymer molecules of a specific size and the molecular weight of the said fraction of molecules; and overall weight $W = \Sigma W_i = \Sigma N_i M_i$.

Weight Average

Weight average molecular weight $\bar{M}_w$ is expressed by the summation of the product of the weight fractions of polymer chains of specified chain lengths and their corresponding molecular weights[2], as shown below:

$$\bar{M}_w = \frac{\Sigma N_i M_i^2}{\Sigma N_i M_i} = (\Sigma W_i M_i)/W$$

$$= \frac{W_1}{W} M_1 + \frac{W_2}{W} M_2 + \frac{W_3}{W} M_3 + \ldots \frac{W_i}{W} M_i + \ldots \qquad (1.4)$$

The ratio of $\bar{M}_w / \bar{M}_n$ gives an idea of the polydispersity of the polymer in question; this ratio is commonly known as the *distribution ratio* or *polydispersity index*. For a theoretically monodisperse polymer $\bar{M}_w / \bar{M}_n = 1$, i.e. $\bar{M}_w = \bar{M}_n$; for most real polymers, $\bar{M}_w / \bar{M}_n > 1$, i.e. $\bar{M}_w > \bar{M}_n$. The polydispersity is higher for a higher value of $\bar{M}_w / \bar{M}_n$.

Viscosity Average

A third average molecular weight, $\bar{M}_v$, called viscosity average molecular weight is conceivable and is expressed as follows:

$$\bar{M}_v = \left(\frac{\Sigma W_i M_i^a}{\Sigma W_i}\right)^{1/a} = \left(\frac{\Sigma N_i M_i^{(a+1)}}{\Sigma N_i M_i}\right)^{1/a} \qquad (1.5)$$

where a is a constant depending upon the polymer-solvent system. For $a = 1$, $\bar{M}_v = \bar{M}_w$ and the viscosity average reduces to the weight average.

Among the important physical and mechanical properties that vary with chain length, i.e. DP or molecular weight, are: *softening point* and *melt flow index (MFI)*, *melt viscosity* and *solution viscosity*, *tensile strength* and *compressive strength*, *elastic moduli*, *impact strength* and *solubility*. Properties such as *refractive index*, *hardness*, *colour*, *density* and *electrical properties* change very little with variation in molecular weight. Many of the polymer properties that have a bearing on molecular weight are highly relevant in adhesives and coatings technology.

1.7 Aspects of Polymer Solubility

High polymers are governed by the same basic laws as applicable to small molecules. However, in view of their large molecular size and the fact that they may be made to undergo drastic changes in shape and dimension, much of their behaviour differs significantly from those of low molecular weight materials. So, manipulation of polymers in melt or solution conditions is altogether a different proposition and experience.

While considering the prediction of solubilities the following points assume importance:

(a) 'Like dissolves like' is the common experience; the aromatic compound aniline is more soluble in benzene, the simplest aromatic hydrocarbon than in the comparable aliphatic hydrocarbon, *n*-hexane. Similarly, *n*-hexane is fully miscible with its immediate higher homologue, *n*-heptane
(b) Comparable polarity induces miscibility and solubility
(c) Solubility is usually higher and dissolution takes place more readily at a higher temperature
(d) Solubility decreases as the molecular weight of the polymer solute of a given type increases

It is to be further noted that solubility relations of a high polymer system appear more complex in view of their complex molecular size and shape, exhibition of high viscosities in solution and the odd distribution and variations in crystalline and amorphous regions in them in the fibre form or in the glassy and rubbery states, which again depend largely on the thermomechanical history through which the polymer system is made to pass.

1.8 Factors that Influence and Control the Swelling and Solubility of Polymers

For a solvent to dissolve a mass of solute, the molecules of the solvent must be inherently able to penetrate the solute molecules; they must overcome the cohesive forces between the solute molecules sufficiently to enable their separation from one another and, in effect, under the influence of the overpowering solute–solvent interaction, force them to diffuse into solution. For high polymer solute molecules and simple, small solvent molecules, the solvent penetration into the solute system is usually a slow process, often aided by heat and agitation. The final state of uniform dispersion or dissolution follows a primary state of the solvent system swelling the solute mass. If the specific solvent-solute affinity is high, solvation, progressive swelling and complete dispersion may occur, resulting in a solution, of relatively high viscosity. Dissolution preceded by notable to heavy swelling is evidence that the solute material is a high polymer.

1.8.1 Effects of Polarity, Branching, Cross-linking and Crystallinity

Polymers bearing polar, hydrophilic groups such as —SO_3H, >C=O, —CHO, —COOH, —OH, —$CONH_2$, etc. may swell and finally dissolve in polar solvents such as water, alcohol-water mixtures, formamide, etc. Cross-linked polymers, polar or non-polar, fail to dissolve in a solvent, even though they may show varied degrees of swelling in different solvents, polar or non-polar. Branching imparts a change in the shape of polymer molecules, while cross-linking changes both shape and size. Extensive cross-linking may limit or even prevent swelling. The presence of extensive hydrogen bonding and a relatively high degree of crystallinity in an apparently linear polymer, such as in cellulose, render it insoluble in water and in most common solvents. Non-polar linear and branched polymers are usually soluble in comparable non-polar solvents. Under a comparable environment, the solution viscosity at a given concentration and temperature is higher for a given polymer having a higher molecular weight[2]. A branched polymer is usually more readily soluble in a given solvent than the corresponding linear polymer. Branching lowers the structural symmetry and enhances molecular mobility and chain flexibility. Likewise, copolymerization using a comonomer also produces a copolymer characterized by a poorer molecular symmetry and higher chain flexibility, making the copolymer more readily soluble, too. Branching and copolymerization both lead to lowered structural symmetry in a polymer, thus lowering its density and proneness to crystallization. Consequently, this enhances the ease of solvent penetration, swelling and dissolution. This concept has a strong bearing in the selection of solvents, softeners, plasticizers and other additives in adhesive and coating technology.

1.8.2 Use of Mixed Solvents

The use of mixed solvents appears more effective in certain cases. Partially nitrated cellulose is an important case in point. Part nitration of cellulose lowers its high order of structural regularity or symmetry through part substitution of the —OH groups by —ONO_2 groups, thereby contributing to a notable lowering in the degree of crystallinity and in the scope for intermolecular hydrogen bond formation; a mixture of alcohol and ether proves to be a more effective solvent system and swelling agent for partially-nitrated cellulose than either solvent alone. It is logical to presume that the alcohol exerts the desired solvating action on the residual —OH groups of cellulose nitrate while the ether does so on the nitrate (—ONO_2) groups. Acetone acts as an effective solvating agent for the nitrate groups only of the partially-nitrated cellulose; it is no wonder that, fully nitrated cellulose, viz. cellulose trinitrate readily dissolves in acetone, while it is a relatively poor solvent for partially-nitrated cellulose.

Further, it is interesting to note that having comparable polarity and carbon-hydroxyl ratio for cellulose —$(C_6H_{10}O_5)$—$_n$ and poly (vinyl alcohol) —(C_2H_4O)—$_n$, the latter is soluble in water because of its overtly simple, linear structure and flexibility, while the former, appearing as a polymer of β-glucose rings, is moderately crystalline, significantly rigid and insoluble in water and other common solvents. Under comparable situations, a more flexible polymer exemplified by the presence of —O— , —S— linkages along the chain axis is more readily soluble and deformable. Polar groups such as $—SO_2—$ and cyclic or ring structures in a polymer, especially if they appear in the repeat units along the main chain axis, make the polymer in question more rigid, stiff, heat resistant and solvent resistant, too, as in Table 1.1.

Despite having a high order of crystallinity and solvent resistance, polyethylene $—(CH_2—CH_2)_n—$ is far less rigid and thermally resistant than amorphous polystyrene $-\left(CH_2-\underset{\displaystyle C_6H_5}{\underset{|}{CH}}\right)_n-$; poly (para phenylene) (PPP)

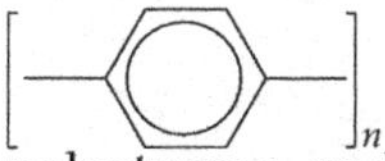

is even more rigid, stiff and resistant to heat and solvents than polystyrene, and PPP, by and large, behaves as a liquid crystalline polymer that is processable only with difficulty.

Rigid, bulky aromatic ring substituents tend to make a vinyl polymer more rigid and brittle. When aromatic rings constitute the chain backbone, the polymer becomes more rigid, stiff, strong and tough; —S— and — O— inter-unit linkages make the polymer weak, relatively soft and flexible, while highly polar $— SO_2—$ linkages along the chain backbone render the polymer more rigid, strong and thermostable.

In case a chain polymer solute containing polar units of one or more kinds in its structure is put into a solvent also containing polar groups, the polar units of the solvent molecules will be oriented toward the polar groups of the chain molecules and the non-polar units of the solvent molecules get directed outward. In effect then, the solvated molecular chains will provide a weaker polar outer surface than the initial polymer. The addition of a second, less polar solvent then goes to enhance the solubility as the second solvent reduces the polarity of the environment to a level closer to that of the surface regions of the initially solvated chains, thereby setting the stage ready for the solvated chains to progressively go into solution. This may also go to partly explain the observed enhanced solvent action of an alcohol-ether mixture on partially-nitrated cellulose[3].

1.9 Solution Viscosity

The measured hydrodynamic properties reflect the behaviour of individual chain molecules only at infinite dilution, such that experimental data obtained at finite concentrations must be extrapolated in general to vanishing or zero concentration. This applies equally to measurements of *osmotic pressure* (for determination of $\bar{M}_n$), *light scattering or turbidity* (for determination of $\bar{M}_w$) and *solution viscosity* (for determination of $\bar{M}_v$) for the assessment of molecular weights and unperturbed dimensions of the chain molecules. In the case of viscosity measurements, this necessity is infused in the very definition of *intrinsic viscosity* $[\eta]$ which is the quantity usually measured or determined. In fact, the definition of $[\eta]$ and its relationship with $\bar{M}_v$ may be given by

$$\underset{c\to 0}{\text{Lt}}\ (\eta_{SP}/c) = [\eta] = K\bar{M}_v^a \tag{1.6}$$

TABLE 1.1 Properties and application prospects of some aliphatic and aromatic (thermoplastic) polymers with the structural identity of repeat units

Polymer	*Repeat Unit*	T_g, °C	T_m, °C	*Remarks*
Polyethylene (linear), PE	$—CH_2—CH_2—$	–115, –60	137	General purpose, low modulus polymer with a good degree of toughness; excellent electrical insulator
Polystyrene, PS	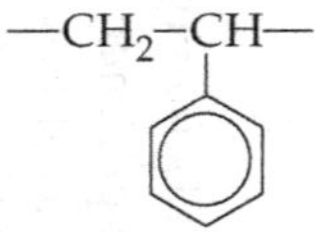	80–100	240	Rigid, brittle, transparent polymer; excellent insulator
Poly (vinyl naphthalene) PVN	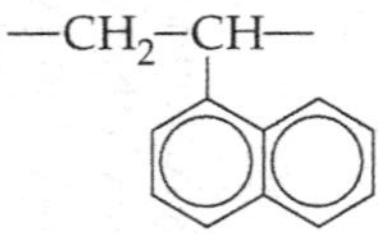	150	360	Highly rigid, brittle polymer; excellent insulator
Poly (paraphenylene), PPP	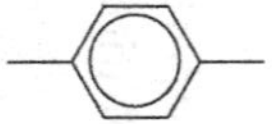	...	530	Tough, thermostable, intractable polymer with semi-conducting or good conducting (when doped) properties with electrical conductivity in the range 10^{-18} to 10^2 $(\Omega\ cm)^{-1}$
Poly (tetramethylene oxide), PTMO	$—(CH_2)_4—O—$	–80	60	Weak flexible polymer
Poly (phenylene ether), PPE	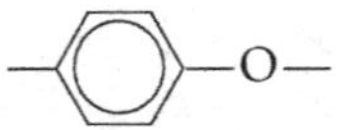	80	300	A good engineering thermoplastic with high strength and toughness

(Contd.)

Table 1.1 (*Contd.*)

Poly (phenylene oxide), PPO	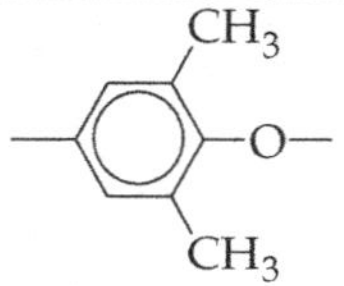	209	...	Very low water absorption, more resistant (chemically, mechanically and thermally) than PPE
Poly (tetramethylene sulphide), PTMS	$—(CH_2)_4—S—$	...	70	Weak, flexible polymer
Poly (phenylene sulphide), PPS	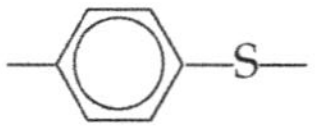	90	295	Strong, tough, thermally resistant polymer. Normally a good insulator, but when properly doped, say with 2-5% AsF_5, it assumes a semiconducting character
Poly (tetramethylene sulphone), PTMSO	$—(CH_2)_4—SO_2—$	...	270	The presence of strongly polar sulphone ($—SO_2—$) links makes the polymer mechanically strong and thermally more stable, as can be realized by comparing PTMSO with PTMS or PTMO
Poly (phenylene sulphone), PPSO	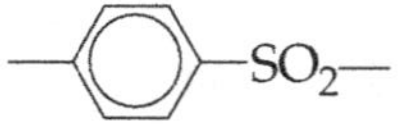	350	525	Mechanically very strong and thermally stable upto 500°C; high-performance plastic
Poly (ether sulphone), PES	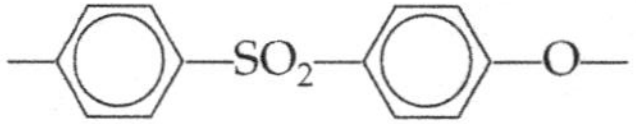	230	...	Thermostable, high performance, transparent thermoplastic with improved processability compared to PPSO
Poly (ether ketone), PEK	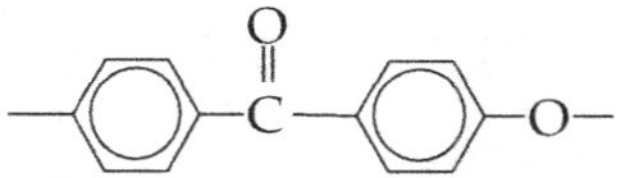	155	365	High performance thermoplastic with excellent environmental stress cracking resistance; continuous use temperature nearly 275°C
Poly (ether ether ketone) PEEK	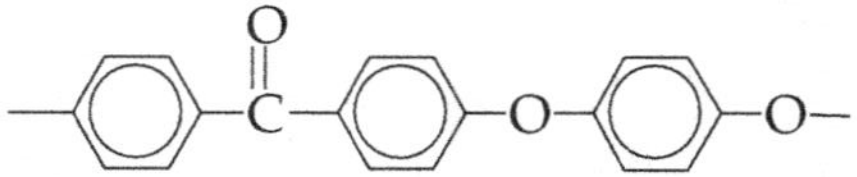	140	343	High performance, chemical resistant thermoplastic, showing a high, continuous use temperature (250°C)

The *specific viscosity* η_{SP} at a concentration c is given by $(\eta - \eta_o)/\eta_o$ or $(\eta/\eta_o) - 1$, where η is the viscosity of the solution and η_o is the viscosity of the solvent at a given temperature of measurement; (η/η_o) is the *relative viscosity* (dimensionless) and (η_{SP}/c) is known as the *reduced viscosity*, bearing the unit of reciprocal concentration. The concentration c is usually expressed in g/dl and the unit of $[\eta]$, giving a measure of the hydrodynamic volume intrinsic under the condition of infinite dilution, is dl/g. (η_{SP}/c) may be strongly concentration dependent for even dilute polymer solutions, often expressed[4] as:

$$(\eta_{SP}/c) = [\eta] + k\,[\eta]^2\,c \tag{1.7}$$

where k is the Huggins' constant[4]. The intrinsic viscosity $[\eta]$ is also given by an alternative equation known as Kraemer's equation[5]:

$$(\ln \eta_r)/c = [\eta] + k'\,[\eta]^2\,c \tag{1.8}$$

or

$$\underset{c \to 0}{\mathrm{Lt}}\ (\ln \eta_r)/c = [\eta] = K\,\bar{M}_v^a \tag{1.9}$$

The parameter $(\ln \eta_r)/c$ is commonly referred to as the *inherent viscosity*, k' as Kraemer's constant and the parameter $[\eta]$ is also called, *the limiting inherent viscosity* or *the limiting viscosity number.*

Equations (1.7) and (1.8) permit plots of the reduced viscosity, (η_{SP}/c) vs c and inherent viscosity, $[(\ln \eta_r)/c]$ vs c using a common ordinate and abscissa, as in Fig. 1.3. The two plots give a common intercept on the ordinate, thus enabling the precise determination[6] of intrinsic viscosity, $[\eta]$. For most cases, k' is negative and in general $k + k' = 0.5$. The slope of each plot is proportional to the square of the intercept, i.e. to $[\eta]^2$.

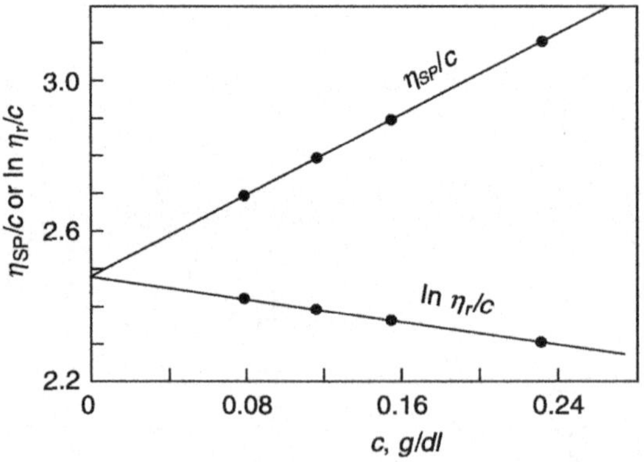

FIGURE 1.3 Typical plots of (η_{SP}/c) vs c and $(\ln \eta_r)/c$ vs c at a fixed temperature for a given polymer-solvent system for the determination of intrinsic viscosity $[\eta]$ from the (common) intercept.

The viscosity η of a polymer solution is higher than the viscosity η_o of the pure solvent at a given temperature; the gain in medium viscosity on dissolving the polymer in the solvent is a function of both molecular weight and concentration of the polymeric solute. If the polymer solution is very dilute and, consequently, the density change of the solvent due to dissolved polymer is negligible, then the viscosities of the solution and solvent at a given temperature would be proportional to their flow times in a capillary viscometer[2,6] (Ostwald or Ubbelohde types, see Fig. 1.4) such that the relative viscosity η_r expressed as the ratio η/η_o would be given by the flow time ratio, t/t_o, where t and t_o are the flow times, usually in seconds, of the solution and solvent respectively.

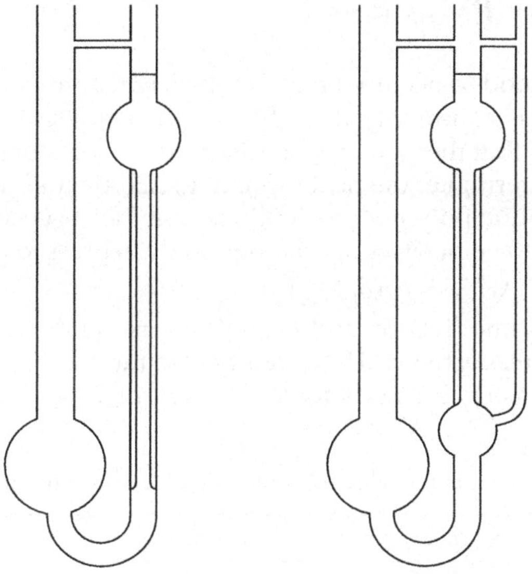

FIGURE 1.4 Capillary viscometers: (a) Ostwald and (b) Ubbelohde types

1.10 OTHER CONSIDERATIONS

The presence or absence of solubility, as the environmental conditions (such as the temperature, stirring or agitation and the nature of the solvent, including selected solvent mixtures) are varied, can give a range of useful information about the polymer.

The absence of solubility does not necessarily imply the presence of cross-linking. Other features may contribute to sufficiently high intermolecular attractive forces, both intrinsic between neighbouring chain segments and cumulative over the entire chain molecules, so as to prevent solubility. The presence of a high degree of crystallinity consequent to extensive intermolecular/intramolecular H-bonding is a matter of special importance in this context. Crystalline polymers, especially the non-polar ones, do not dissolve except at temperatures near their crystalline melting points (T_m) because crystallinity decreases as T_m is approached and also because the melting point itself is lowered by the presence of the solvent. Hence, solubility can be variously achieved at temperatures measurably below the melting point.

Thus, linear polyethylene, the well-known high density polyethylene (HDPE), having crystalline melting point T_m = 133°C, is actually soluble in toluene or xylene or xylene-toluene mixtures at or near 100°C. Likewise, the low density polyethylene (LDPE), having a lower degree of crystallinity and the crystalline melting point $T_m = 117°C$, can be somewhat readily dissolved in the said solvent or solvent mixtures at or near a much lower temperature, 80°C.

1.11 Solubility Parameter

Solubility is understood to occur when the *free energy* of mixing ΔG (= $\Delta H - T\Delta S$) is negative. It is commonly thought that the *entropy* of mixing ΔS is always positive such that the sign of ΔG is determined by the sign and magnitude of the *enthalpy* term, i.e. the heat of mixing, ΔH. By and large, for non-polar solvent and solute polymer molecules and in the absence of hydrogen bonding, ΔH is said to be positive and is commonly expressed as:

$$\Delta H = v_1 v_2 (\delta_1 - \delta_2)^2 \tag{1.10}$$

where v denotes volume fraction and the subscripts $_1$ and $_2$ refer to solvent and polymer solute respectively. The quantity δ^2 is the *cohesive energy density* and, for small molecules, it stands for the energy of vaporization per unit

TABLE 1.2 Solubility parameter values of some monomers/solvents and polymers[8,9]

Monomer/Solvent	δ_1 $(cal/cm^3)^{1/2}$	*Polymer*	δ_2 $(cal/cm^3)^{1/2}$
Methanol	14.5	Cellulose	15.65
Ethylene carbonate	14.7	Cellulose diacetate	11.35
2-Ethyl hexanol	9.5	Cellulose nitrate	10.72
Dioxane	9.9	Poly (methyl acrylate)	10.15
n-Butyl acetate	8.5	Polyacrylonitrile	12.50
Acetonitrile	11.8	Polychloroprene	9.38
Toluene	8.9	Poly (ethylene terephthalate)	10.7
n-Butanol	11.4	Polymethacrylonitrile	10.7
Acetone	9.9	Poly (vinyl chloride)	9.42
Benzene	9.2	Poly (vinylidene chloride)	12.2
Ethyl alcohol	12.7	Poly (vinyl acetate)	9.59
Methacrylic acid	11.2	Poly (vinyl alcohol)	12.60
Acrylic acid	12.0	Polystyrene	8.56
Quinoline	10.8	Polyethylene	7.70
Pyridine	10.7	Polypropylene	9.20
Caprolactam	12.7	Polyisobutylene	7.85
Cyclohexanone	9.9	Nylon 66	13.60
Chloroform	9.3	Natural rubber	7.9
Chloro benzene	9.5	Poly (methyl cyanoacrylate)	14.0
Acetic acid	10.1	Alkyd resin (medium oil length)	9.4
Ammonia	16.3	Poly (dimethyl siloxane)	7.35
Glycerol	16.5	Epoxy resin	10.9
Formamide	19.2	Benzyl cellulose	12.33
Water	23.4	Poly (butadiene-co-vinyl pyridine)	9.35

volume. The *quantity δ is called the solubility parameter*[7]. Miscibility or solubility can be expected if δ_1 and δ_2 values are close and $(\delta_1 - \delta_2)$ is less than 1.8; but solubility is not favoured if the difference in the two solubility parameters is much higher, particularly in systems devoid of strong interactions such as hydrogen bonding. The typical values of solubility parameters of some solvents and polymers are listed in Table 1.2.

1.12 Solvent Power and Good Solvents

With the thermodynamics of the dissolution of polymers as above and having some exceptions as well in that both ΔH and ΔS can be negative, the applicability of the solubility parameter concept may at times appear invalid or inappropriate. By and large, however, the concept is useful even after allowing for the mismatch for the handful of unusual cases. The rate of dissolution primarily depends on how rapidly the polymer-solvent system would diffuse into one another. Kinetically good solvents are not necessarily good thermodynamically. However, mixtures of kinetically and thermodynamically good liquids often prove to be powerful and quick dissolving solvents for polymers.

It is a common characteristic of high polymers that at concentrations of only a few percent, their solutions are very viscous or may even form very thick, non-flowable gels. Spherical molecules or particles impart lower resistance to flow than oblong or elongated ones. An elongated particle rotates in solution and covers an effective volume greater than its actual volume. This is unlike particles that are spheres, whose actual and effective volumes are equal or comparable.

It follows therefore, that in a good solvent the chain molecules would be in a more expanded or elongated condition than in a poor solvent. The intrinsic viscosity (a measure of the hydrodynamic volume of a unit mass of the selected polymer solute) in a good solvent is higher than in a poor solvent. At progressively high concentrations however, the molecular chains will tend to associate or cluster into larger aggregates. Such trends of association will be more severe in the poor solvent than in the good one, particularly beyond a low threshold concentration in each case, as in Fig. 1.5. It is seen clearly from the trends of change in reduced viscosity with increase in concentration *c* that the sharp upward turn appears early in the concentration scale for the poor solvent, and that the two plots intersect or cross each other at a low intermediate concentration. The rate of increase in viscosity with concentration is decidedly greater in the poor solvent. It is no wonder then that a good solvent (giving a higher intrinsic viscosity) not only dissolves more polymers but also produces solutions of lower (reduced) viscosities at somewhat intermediate or higher concentrations.

These features are of immense theoretical and practical significance, particularly with respect to solvent-based coatings. In designing and formulating

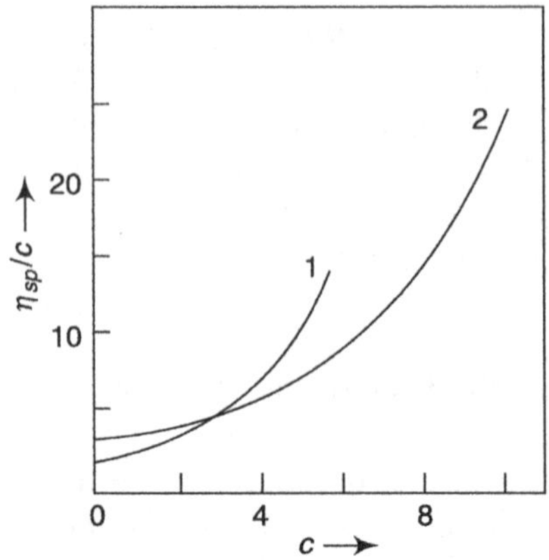

FIGURE 1.5 Trends of change of (η_{SP}/c) vs c in g/dl for: (1) a polymer solution in a poor solvent and (2) a solution of the same polymer in a good solvent

a lacquer, it is instructive to have as low a viscosity and as high a concentration as possible at the same time. A high concentration would enable the formation of a film of suitable thickness for one or fewer applications, while a low viscosity would enable easy, trouble-free application and spreading of the solution by brushing or spraying. The most appropriate choice of a solvent system is thus of high significance and great consequence.

The utility and significance of mixed solvents have already been highlighted (1.8.2) with respect to cellulose nitrate-based lacquers. Viscosity measurements of solutions of lacquer-grade cellulose nitrate (DS = 1.9–2.2) in some ester (ethyl acetate or butyl acetate)-ethyl alcohol mixed solvents indicate that progressive additions of alcohol up to 50% to 60% by volume enhance the solubility of the polymer and reduce the solution's viscosity. Further additions of alcohol reduce the solubility and increase the viscosity again.

Solvent power may also be assessed by the dilution ratio approach: the high polymer solute is dissolved to a given standard strength in each of a number of selected solvents. One then needs to determine the amount of a specific non-solvent required to be added, on stirring, to cause the appearance of light, permanent turbidity or initial precipitation. The solvent to which the highest amount of the non-solvent is needed to cause the first turbidity or precipitation to appear, is the most powerful.

The dilution-ratio approach suffers from the weakness that the initial additions of the diluent or non-solvent may in fact improve the solvent power of the (mixed) solvent system, much in tune with the effects already outlined. Even then, the approach is of great practical relevance with regard to solvent balance and striking a better economy through the replacement of costly solvents by less expensive or more readily available diluents or latent solvents.

1.13 THIXOTROPY

The development of primary valence cross links in an initial polymer remaining in solution commonly turns it into an irreversible gel due to the ultimate formation of a giant macromolecular network, thus causing the immobilization of the *solvated* and swollen polymer mass; once formed, the gel-mass cannot be redissolved by thermal or mechanical energy, i.e. heat or stirring. The gel so formed may be broken by stirring it into smaller, swollen pieces; it can be degraded and thermally decomposed with drastic treatments but can seldom be redissolved to uniformity. This kind of gelation is thus irreversible.

In many other instances, a high polymer solution of a relatively low viscosity may set to a thick gel when left undisturbed at a given temperature. The gel so formed is readily broken simply by agitation or stirring, but it gets readily reformed on standing again. This unique reversible isothermal sol-gel transformation phenomenon is commonly called *thixotropy*. It should be mentioned that many such gels show a slow trend to densify on long standing with associated sweating or squeezing out of the solvent as a separate phase, a phenomenon called *syneresis*. A thixotropic gel can very well be broken merely by adding a controlled proportion of a second solvent of the right character. Again, a considerable portion of a thixotropic gel mass can possibly be leached out in selected cases without actually breaking the gel structure.

A thixotropic gel is evidently a weak, continuous, dendritic kind of structure held in place by a massive interplay of secondary valence bridging, and it is altogether different from a space-polymerized network gel. Thixtropic gel formation is favoured in a solution of a basically non-polar polymer bearing widely separated polar groups in its structure in a non-polar solvent. Through polar-polar interactions (mostly intermolecular and partly intramolecular) between the widely-spaced polar groups, a continuous, dendritic structure build-up with the entrapment of the solvent mass at least in part will occur and the whole mass will eventually get immobilized into a gel. The gel structure so formed is weak and readily broken, causing the mass to assume the character of a more or less free-flowing viscous solution again, simply by stirring or shaking. This process of gel formation on standing and the break-down of the gel to a flowing liquid mass on mechanical agitation at the same specified temperature may be repeated several times. Thixotropic gels are complex in nature and their responses to the addition of solvents in different volume levels and of different polarities are also quite complex. The thixotropic effect does not necessarily cause gel formation, which may be an extreme manifestation; it may simply involve a physical change from a less viscous solution/ liquid mass on stirring to a relatively thick, more viscous liquid/solution mass on standing.

Rubber, ordinarily a non-polar natural hydrocarbon polymer (1, 4 cis-poly isoprene), is masticated to generate adequate tackiness in the first stage of its processing, during which more polar groups are generated on the degraded rubber chains through oxidation. Rubber adhesives/cements are made by dissolving the masticated rubber in a non-polar hydrocarbon such as solvent naphtha. Even on mastication, it is difficult to obtain a workable (spreadable/ brushable) solution of concentration more than a few percent; a higher concentration leads to the undesirable formation of a gel-type mass or a solution of very high viscosity. To dissolve more rubber and at the same time keep the viscosity low for trouble-free manipulation and application, it is advantageous to add an adequate amount of a polar solvent such as alcohol, which lowers the trend of expansion of the rubber hydrocarbon chains in solution and restricts the swelling, gelling or thixotropic effect.

1.14 Advantages of Thixotropy in Adhesives and Coatings

Thixotropic effects are exhibited by a large array of colloidal and high polymeric systems such as lacquers, varnishes and paints, greases, gums and starch solutions, clay suspensions, inks, gelatin; many protein solutions, plant juices and other biological fluids; emulsions of different kinds; bitumens, greases and lubricating oils at low temperatures. In fact, thixotropy is highly desirable or even essential in liquid surface coatings, adhesives and greases.

In a good paint, the consistency/viscosity must be high enough to prevent the settling of pigment on standing. Through a process called caking, the settled pigment forms a compact, hard mass at the bottom that is far too difficult to be stirred back to homogeneity. At the same time, the consistency should be low enough to enable brushing to a coat of uniform thickness; at the same time, brush marks must quickly disappear or flatten out by flowing before the paint dries up. On the other hand the initial consistency as well as its regain after application must be high enough to prevent the brushed or sprayed paint from running down after it is laid on a vertical surface, and prevent a sag or drip-off of the paint from the bottom side of horizontal surfaces. A fine balance of rheological properties is necessary to infuse all these conflicting requirements into the paint mass. Only a precisely designed thixotropic paint system can offer such conflicting yet balanced requirements. Caking troubles can be avoided or minimized only if the paint is thixotropic. During application, under the shearing action of stirring followed by brushing or spraying the coating material on a substrate surface, the viscosity drops considerably, thus permitting easy flow and spread. The rate of thixotropic set or recovery to a gel state from the spread out paint to a film of desired coverage and thickness occurs rapidly enough to prevent its running down the vertical surfaces or sagging down the bottom side of horizontal surfaces.

1.15 Mechanistics of Thixotropy

The high consistency of polymer solutions or polymer dispersions/emulsions is drastically lowered at high shear rates, and regained on standing thereafter, thus indicating its thixotropic character. The effect may be understood by assuming that surface forces cause the particles at rest to turn into filament-like, clustered chains, which then form a mesh-like network structure by continuous scaffolding, finally giving a strong, high gel consistency. Sustained mechanical agitation or stirring supplies enough energy to break down the scaffolding or gel network built up by the interplay of secondary valance bridging; the consistency then drops to a normally expected value. Set at rest, the scaffolding of the growing filamentous chains of the particles is reformed to full potential and the apparent viscosity rises again.

REFERENCES

1. Staudinger, H, Ber, Vol. 53B, 1920, 1073.
2. Ghosh, P, *Polymer Science and Technology—Plastics, Rubbers, Blends and Composites,* 2/E, Tata McGraw-Hill, 2002.
3. Schmidt, A X and C A Marlies, *Principles of High Polymer Theory and Practice,* McGraw-Hill, 1948.
4. Huggins, M L, *J. Amer. Chem. Soc,* Vol. 64, 1942, 2716–2718.
5. Kraemer, E O, *Ind. Eng. Chem.,* Vol. 30, 1938, 1200–1203.
6. Billmeyer, Jr, F W, *Text Book of Polymer Science,* 3/E, Wiley-Interscience, 1984.
7. Small, P A, *J. Appl., Chem.,* Vol. 3, 1953, 71–80.
8. Walker, E E, *J. Appl. Chem.,* Vol. 2, 1952, 470–481.
9. Brandrup, J and E H Immergut, eds, *Polymer Hand Book,* Vol. IV Interscience, 1967, 345-367.

Chapter

2

Polymer Rheology

A body in any physical form gets deformed when subjected to an external stress. The trend of a body to resist deformation is called its *consistency*. Deformation may be of two basic types: (a) (Hookian) *elastic deformation,* which appears temporarily on stress application and disappears on its stress withdrawal and (b) (viscous) *flow,* which is manifested as a permanent deformation. *Rheology* concerns the deformation and flow of matter. Rheological studies relate the mechanical behaviour of a material with respect to three variables: *stress, strain* and *time.*

In the event of a permanent deformation, i.e. flow occurring regardless of whether the body is pulled, stretched, bent, compressed, twisted, dented or torn, the common meaning is that particles within the body have permanently slipped past one another or some of them have slipped over each other under shear, overcoming the restraining influence of inter-particle forces of attraction.

2.1 Ideal Elastic Deformation

An *ideal* elastic deformation occurs instantaneously in what is called a Hookian solid, on applying a deforming stress and the deformation is only temporary as it disappears completely and instantaneously on release of the deforming stress, which is within the elastic limit. Here, the strain developed is independent of time and a function of the applied stress alone.

By mechanical analogy, the rheological behaviour of a Hookian, i.e. *ideally elastic body* (that follows Hooke's law as in Eq. (2.1)) is that of a steel spring. Hooke's law states that instantaneous *strain* (γ) developed on an ideally elastic body at a given temperature is proportional to the *stress* (τ) applied such that:

$$\frac{\text{Stress}(\tau)}{\text{Strain}(\gamma)} = E \qquad (2.1)$$

where the proportionality constant E is known as the *modulus of elasticity* or *Young's modulus*.

Stress $\tau = (F/A)$, where F is the force applied on an area A and the developed *strain*, $\gamma = (\Delta L/L_0)$, where ΔL is the change in length or dimension and L_0 is the initial length/dimension.

For a Hookian body under tension or compression, the modulus is referred to as Young's modulus while when it is under shear, the *shear modulus* $G = (\tau_s/\gamma_s)$ is referred to as *modulus of rigidity*[1].

Elongation under tensile stress is concomitantly associated with a lateral contraction and ratio of lateral to longitudinal strain, called Poisson's Ratio is expressed by the symbol v; the value of v is equal to, greater or less than 0.5 for density of the material remaining unchanged, increasing or decreasing on deformation under stress[1]. Typical stress-strain plots for some real bodies (thermoplastic and rubbery) are shown in Fig. 2.1.

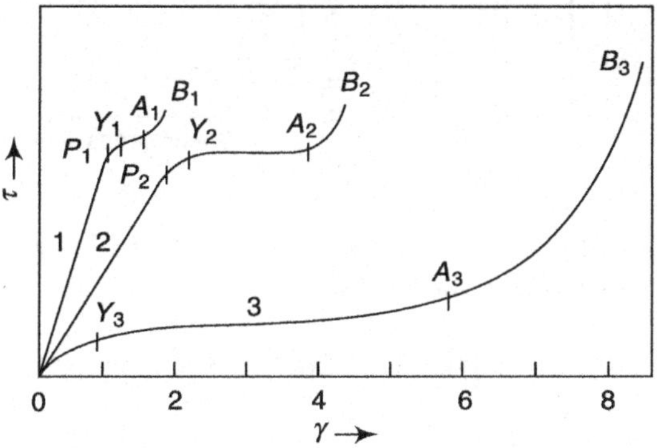

FIGURE 2.1 Stress (τ)–strain (γ) plots of typical thermoplastic materials (1 and 2) and a typical rubber (3): (Curve 1) is for a stiff and rigid thermoplastic, (Curve 2) is for a strong and tough thermoplastic and (Curve 3) is for a vulcanized rubber; O: origin; P_1, P_2: elastic limits; OP_1, OP_2: Hookian zones; Y_1, Y_2, Y_3: yield points; Y_1A_1, Y_2A_2 and Y_3A_3: strain hardening zones; B_1, B_2 and B_3: breaking points (schematic)

2.2 Ideal or Newtonian Fluid

When a fixed or constant stress is applied to a liquid or fluid body, it undergoes a continuously increasing amount of strain or deformation, which is permanent and non-recoverable on withdrawal of the stress. A liquid is thus a material of zero yield value in which the strain (γ_s) is not only a function of stress (τ_s) but also of the time (t) of stress application, i.e.

$$\gamma_s = \frac{\tau_s}{\eta} \cdot t \tag{2.2}$$

Here, the coefficient of viscosity η is a true constant of the fluid body, independent of the shear stress and shear rate or rate of flow (without turbulence) at a given temperature; η, the viscosity of the fluid body, does not depend on its prior shear deformation history. A plot of shear rate $(d\gamma_s/dt)$ vs shear stress (τ_s), in Fig. 2.2 shows a straight line passing through the origin for which the slope ϕ represents *fluidity* or the reciprocal of *viscosity*[1].

A *Newtonian liquid*, totally devoid of *elasticity*, is viewed as a *viscous* material and, by mechanical analogy, its deformation manifested as *flow* under stress is represented by a weightless piston moving in a cylinder or dashpot filled with a fluid medium that offers some resistance, as in Fig. 2.2 (a). The amount of flow or strain γ_s is a linear function of time as in Eq. (2.2).

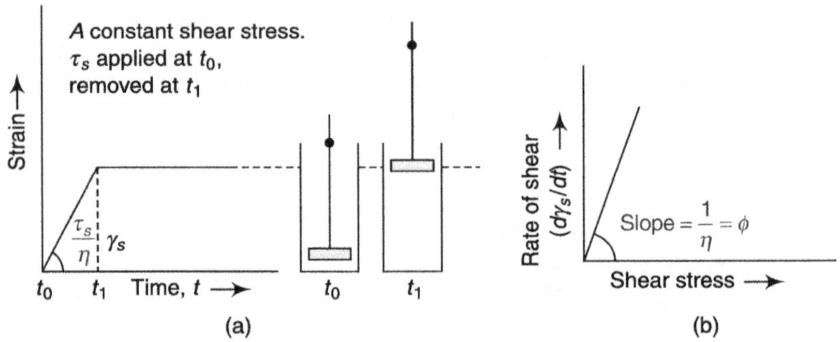

FIGURE 2.2 Flow behaviour of a Newtonian fluid: Plots of (a) strain (γ_s) vs time (t), (b) rate of shear $(d\gamma_s/dt)$ vs shear stress (τ_s)

2.3 Non-Newtonian Fluid (Anomalous Flow Behaviour)

Flow behavioural patterns of a Newtonian fluid and different types of non-Newtonian fluids in the form of flow curves, τ_s vs $(d\gamma_s/dt)$ are shown in Fig. 2.3. The flow pattern of a real fluid is better understood by the graphical comparison of the corresponding shear stress (τ_s) vs shear (strain) rate $(d\gamma_s/dt)$ curve with the straight line plot of (passing through the origin) of a Newtonian (ideal) fluid, using a common axis as in Fig. 2.3. From the flow curves in this figure, one can observe and appreciate how the flow behaviours of some real or non-Newtonian fluids (polymer melts, solutions, emulsions, pastes and dispersions) differ from that of an ideal or Newtonian fluid. A deviation from linearity is a common experience for such plots for non-Newtonian or non-ideal fluids[1,3]; the plots of many such real fluids are either concave upwards or concave downwards or more complex in nature. For most non-Newtonian fluids showing anomalous flow patterns, the *coefficient of viscosity* depends on stress level, strain rate and previous shear history of the sample.

For an ideal plastic called *Bingham Plastic,* a finite strain rate is generated only after the material reaches a definite stress level known as yield stress $(\tau_s)_y$; then the stress-strain rate relationship maintains linearity as for a Newtonian fluid.

The behaviour of most real plastics is, however, characterized by non-linear relationships, which are concave downwards as exemplified by

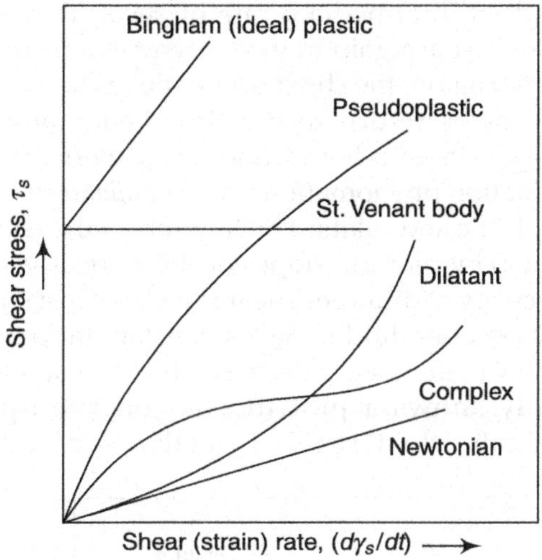

FIGURE 2.3 Flow behaviour patterns [τ_s vs $(d\gamma_s/dt)$ plots] for a Newtonian fluid and several non-Newtonian fluids

pseudoplastic behaviour in Fig. 2.3. Most plastic melts, solutions or dispersions exhibit pseudoplastic flow patterns. At rest, long chain molecules of a plastic melt, solution, emulsion, paste or dispersion are believed to engage themselves into clusters or close stable associations as a result of chain entanglements over and above the interplay of molar cohesion. At relatively low shear rates at a given temperature, the stress required to overcome or undo the effect of *chain entanglements* is of a nature that produces a relatively high slope (indicating correspondingly high viscosity). Stirring at higher rates leads to significant lowering in *viscosity* (or increase in *fluidity*). Increase in shear stress causes more than a proportionate increase in the flow rate, once the molecules are set in motion with respect to each other. This feature is characteristic of pseudoplastic flow behaviour. Finally, at quite high shear rates, the molecular/segmental entanglement effect becomes virtually non-existent and Newtonian or near-Newtonian behaviour assumes prominence, depending on the exact nature of the fluid material being studied. If allowed to stand at rest thereafter, the fluid system may soon *gain* in viscosity and *set* again. Such isothermal sol-gel transformations are commonly known as *thixotropy*. The polymer particles assume a random disposition favouring formation of *scaffolding* or *network clusters* at rest or at low rates of shearing, but they tend to align or orient in the direction of flow as the rate increases and thereby offer substantially lower resistance.

However, an altogether different nature of flow behaviour is characterized by a shear stress (τ_s) vs shear rate ($d\gamma_s/dt$) curve that is concave upward; a fluid of such a flow behaviour pattern is known as *dilatant*. This type of flow behaviour is exemplified by some highly concentrated suspensions of certain fine powders in appropriate liquids and by some melts in rare cases, where crystallization or structure formation occurs at relatively high rates of shearing. Such fluid systems usually resemble a Newtonian fluid at rest or at low shear rates, but beyond an intermediate rate of shear, they sharply form oriented clusters to show a sharp gain in viscosity with a further increase in shear rate. When set at rest again, the shear orientation is no longer sustained and the molecules/particles return to the flow behaviour of a normal (Newtonian) or near-normal viscous liquid. Such an isothermal shear-dependent gel-sol-gel transformation phenomenon is called *dilatancy.*

Newtonian, pseudoplastic and dilatant behaviours of the respective fluids may also be clearly shown by plotting slopes of shear stress vs shear (strain) rate, i.e. of apparent viscosity at different shear rates[1], against shear rate, as in Fig. 2.4. A Newtonian (viscous) fluid shows a constant (apparent) viscosity (straight-line plot) parallel to the shear rate axis, while a dilatant (shear-thickening fluid) commonly shows a plot that is concave upward and a pseudoplastic (shear thinning fluid) shows a plot that is concave downward.

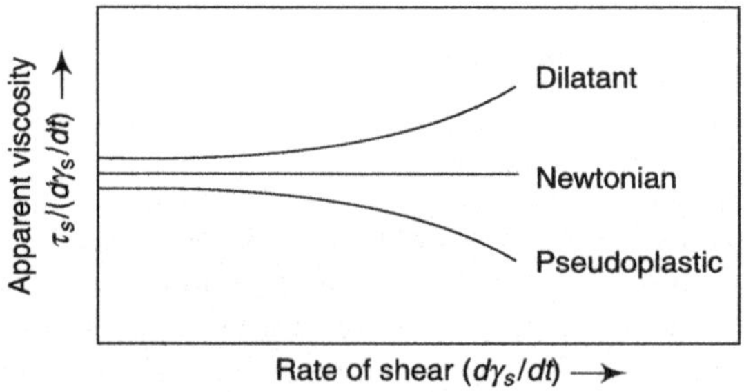

FIGURE 2.4 Plot of apparent viscosity $[\tau_s/(d\gamma_s/dt)]$ vs strain rate $(d\gamma_s/dt)$ for: (1) Newtonian, (2) dilatant and (3) pseudoplastic fluid systems

Dilatants are shear thickening, while pseudoplastics are shear thinning fluid systems. Some systems that form thixotropic gels are endowed with the property of solidifying or thickening more rapidly when sheared (agitated) very slowly than when left at complete rest. Apparently, very slow stirring aids normal diffusion so as to enable the dispersed particles or dissolved chain molecules to arrange themselves into a scaffolding structure. This distinctive property is called *rheopexy*. Ultrasonic vibration may also induce a *rheopectic effect*. Rheopexy is quite different from dilatancy; the equilibrium state for a dilatant is a liquid (sol) state while that for a rheopectic system is a near solid gel. Dilatancy may occur in some coatings, dispersions, filled plastics and paints.

Some systems such as solutions of cellulose esters and ethers, rubbers, boiled starch and other gums, polysachharides or asphaltic bitumens produce non-linear velocity gradient ($D_s = (d\gamma_s/dt)$) vs τ_s plots, as in Fig. 2.5.

The D_s vs τ_s curve of types 1 and 2 is characteristic of pseudoplastic liquids with variable apparent viscosity and absence of a real yield stress. Curve 1 and the initial non-linear portion of Curve 2 can be expressed by the de Waele-Ostwald formula:

$$D = \frac{1}{\eta'} \cdot \tau^n \tag{2.3}$$

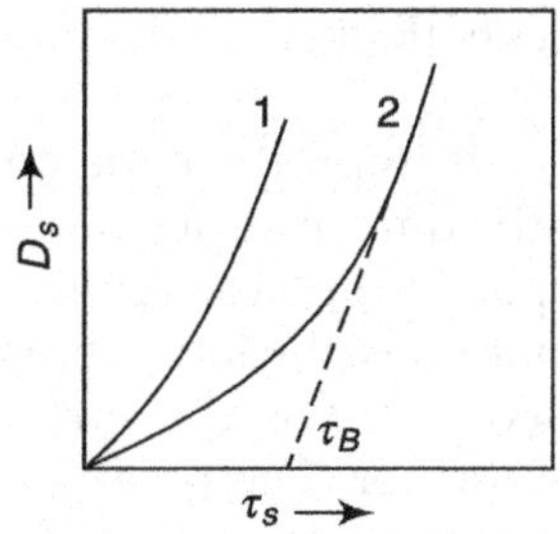

FIGURE 2.5 Non-linear D_s (shear strain rate) vs τ_s (shear stress) plots for two real fluids of different pseudoplastic characters

where the power $n > 1.0$ and the constant η' is not really the viscosity coefficient. For Curve 2, the linear portion at the higher τ_s range is given by:

$$D_s = \frac{1}{\eta}(\tau_S - \tau_B) \tag{2.4}$$

and the constant η now has the dimension of a viscosity coefficient.

One may theoretically encounter a fluid body showing a finite yield stress that would shear at any high rate without requiring the enhancement of stress values above the yield stress. Such a fluid is commonly referred to as a *St. Venant body.*

Rheology is not only useful for the control and study of flow variables in relation to adhesives, plastics, paints, coatings, lubricants, rubbers and many other materials, but also to gain a greater insight into the molecules and interplay of forces that make up matter[1].

2.4 Polymer Dispersions for Adhesive/Coating Applications

2.4.1 Hydrosols

Natural and synthetic latices are the most common application of *hydrosols* in polymer technology. *Hydrosols* are dispersions of solids in water while the term *latex* is commonly reserved for emulsified dispersions of polymers in water. The early Egyptians and Hebrews mixed dispersions of egg white, casein, milk and glue or curds for making paint.

Latex-based paints came into prominence after the butadiene-styrene copolymer latex prepared by emulsion polymerization began to sell in the market. The polymer/copolymer latex is merely a major adhesive or coating ingredient along with many other formulating ingredients such as pigments, plasticizers, sequestering agents, antifoam agents, emulsifiers and protective colloids. The water-soluble protective colloids contribute to enhance the viscosity and stabilization of the dispersed polymer/copolymer particles, on top of rendering the latex prominently shear-thinning, such that it will not drop from the applicator at a low rate of shearing or it will flow readily under application conditions and quickly thicken on the painted surface to prevent sagging.

The stabilizing ingredient needs to produce the much-desired "break" or phase inversion in the latex. At the time of application of the latex coating, the water is in the continuous (dispersion) phase. As evaporation of water from the spread-out, coating layer continues, the latex "breaks" and in the process, the water becomes the dispersed phase on the spread-out, continuous, wet film phase of the polymer/copolymer. The break or phase inversion is required to occur before the polymer dries to a cohesive film but not too soon to adversely affect application and flow-out. With the rapid growth of latex paints/adhesives in volume and variety, the formulator plays a key role in striking an excellent rheological balance. In home and indoor applications, not only easy brushability and proper flow-out, but also non-inflammability and low or no odour are important. If duly crosslinked, a *hydrosol* would turn into a *hydrogel* as exemplified by modern *super sluppers or super soppers* that are fibre (say, cellulose)-anchored or grafted hydrophilic gels having a very high capacity to absorb water and retain it without being dissolved.

2.4.2 Organosols

Organosols are simply fine dispersions of a solid (polymer, resin or others) in an organic liquid. One of the most commercially successful organosols is based on poly (vinyl chloride) (PVC) resins. To start with, the base material is prepared by emulsion polymerization of vinyl chloride monomer and with due control, the emulsion consists of essentially spherical particles in the particle size range of 0.1–1.0 micron in diameter. Next, the actual organosol is prepared by grinding the emulsion resin particles in a ball mill using a mixed organic solvent system as the dispersion medium[4]. The initial high viscosity of the mixture rapidly drops down as milling proceeds to an optimum stage, leading to a break down in the particle size of the emulsion resin agglomerates. Excessive grinding causes the viscosity to go up much as a consequence of over-solvation of the particles by the organic dispersion medium, i.e. a right mixture of a good solvent (dispersant) cyclohexanone and a weak solvent (diluent) in the form of a plasticizer such as disooctyl phthalate.

Too much use of the diluent causes the viscosity to rise, much as a consequence of the particles flocculating and causing grainy, coarse structure for the organosol. Using excessive dispersant may also cause viscosity to rise due to over-solvation of the resin particles, which may even lead ultimately to limited gelation. The point of minimum viscosity is viewed as the ideal balance of diluent and dispersant, as in Fig. 2.6. It is often advantageous to use the organosol on the dispersant-rich side rather than in the diluent-rich side. Having a little over-solvation results in a smooth, shear-thinning dispersion.

The organosol is amenable to application by a number of coating methods to form a film on a substrate; it is then appropriately heated to induce fusion before all or the bulk of the dispersion medium is gone; too rapid heating is avoided to prevent blistering. Organosol coating technology has an edge over

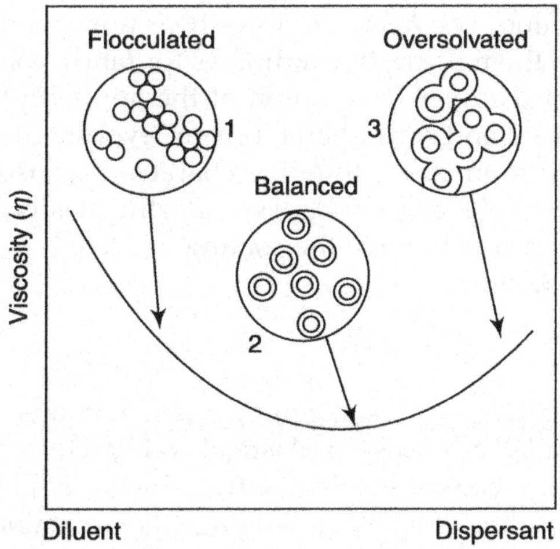

FIGURE 2.6 Viscosity (η) trend and dispersant-diluent balance in organosols: (1) Formation of flocculated clusters of solute particles for high diluent level, (2) moderate (balanced) solvation of solute particles with little trend of cluster formation and adequate dispersion for a good balance of diluent-dispersant ratio, and (3) oversolvation and formation of clusters of swollen mass of solute particles for high dispersant level

solution coating in that (i) the solids concentration of an organosol may be much higher than that of a solution and (ii) an organosol is suited for the manipulation of a notably higher molecular weight resin. Organosols also permit solutions using less expensive solvents and a faster rate of the dispersant/solvent evaporation, as there is less affinity for a solvent towards the polymer in an organosol.

2.4.3 Plastisols

Plastisols are viewed as a special case of organosols. For plastisols, the dispersion medium, in fact a diluent, behaving like a weak solvent, is a high-boiling, non-volatile liquid, commonly called a plasticizer and in such a system, a good volatile solvent is seldom used. Plastisols are commonly prepared in the same manner as organosols, but the dispersion of the customary resin (powder) is merely done by stirring, with no necessity of ball-mill grinding. Enough plasticizer is to be used, in installments, while stirring so as to finally obtain a viscous paste. While mixing, the viscosity developed may initially be very high, but this drops rapidly with time due to the squeezing out of entrapped air in the resin system by plasticizer uptake. On obtaining a smooth paste, more plasticizer may be added as per the flow standard required for a specific processing.

Plastisols may exhibit near Newtonian, pseudoplastic (shear thinning) or dilatant (shear thickening) flow behavioural patterns depending on whether the shearing rate employed is low, moderate or high. Some plastisols may even show a low yield value. The relatively low-shear flow character of appropriate plastisols make them the right candidates for fabric-coating. A low yield value helps prevent the over-penetration of the applied plastisol into the fibres. The coating speed and accompanied shear level should be so controlled as to derive an optimum shear-thinning character to ensure smooth, effective coating. If dilatancy develops at high speeds, the plastisol may actually set up enough resistance to impede the coating or have it scraped from the substrate on roller coating.

2.4.4 Plastigels

A plastisol of a special kind, containing a gelling agent to confer an appreciable yield value on it, is called a *plastigel*. A plastigel is characterized by a high yield value—e.g. modelling clay or glazing putty, which are quite different from heavy pastes. Among the effective gelling agents are aminated bentonite, selected metallic soaps and silica aerogel. Amine treatment of bentonite turns it from an aqueous gelling agent to an organo-gelling agent. When present in an appropriate dose in a plastisol, such gelling agents contribute to some structure formation, thereby resulting in a finite yield point. Such structures are, however, readily broken down on shear, thus enabling the easy formation of objects and shapes by hand pressure, spreading by a knife, extrusion, calendering and moulding. The structure formed may be maintained in some cases even on heating and fusion. In some platisols, only minor doses of some gelling agents are incorporated so as to impart special flow characteristics just to prevent dripping or sagging during fusion.

Plastigels are viewed as a generic derivative of plastisols. Plastigels are associated with some novelty and have infused and generated rheological interests in using gelling agents to manipulate new range of coating items. Plastigels are useful in making floor tiles by cold calendering and fusion and in having cold-embossed coated fabrics, all under the umbrella of plasticized PVC technology.

2.5 Solid Dispersed in Gas

Smokes are colloidal dispersions of solids in gases. However, in most practical coating applications, the polymer is used in relatively large particle sizes kept in a state of dispersion or suspension by the velocity of the gas used as the dispersion medium. Polymer powders of assorted desired size distributions are usually sprayed in the air or in an inert gas (N_2) in the presence of a flame or a stream of heated air to produce sintered coatings on the preheated object or target surface. If the object is not preheated, it has to be heated after coating to fuse the coating to a uniform, homogeneous mass adhering to the object surface as a thin film. The coverage, thickness, adhesion and other qualities of the coating as well as the costs of materials, energy and processing and handling would depend on the size and shape of the object, the nature and temperature of the gas carrying the polymer powders and those of the object to be coated, the melting temperature and flow characteristics of the polymer (thermoplastic) or prepolymer (thermoset), etc. and the recovery and recycling potential of the coating material system.

2.6 Liquid Dispersed in Gas

Ordinarily, a fine dispersion of a liquid in a gas or gas mixture is called a mist or fog. One of the more significant applications of liquids in gas is the technology of spray paints. This technology is also recognized as an art to maintain the proper rheological properties. The liquid (dispersion of polymer in a solvent/diluent or a polymer solution in an appropriate solvent) must have the right viscosity to flow through the array of orifices in the spray gun. The polymers in latex, dispersion or solution form, i.e. in a liquid medium, along with assorted additive ingredients such as pigments and fillers, stabilizers and antioxidants, antistatic agents, blowing agents, etc. are put in a paint spray container and delivered into a paint spray nozzle, where the mix is combined with a high velocity air jet. This causes the liquid to break up into a fine mist or spray ejected by the compressed air onto the substrate to be coated.

For an industrial paint, striking the proper, balanced rheological properties is an art as much as science and technology. The liquid formulation must attain just about the adequate viscosity and flow character to be sprayable into a mist without difficulty or excessive cobwebbing or stringing. On deposition on the substrate surface, it is required to flow out just enough to create a continuous film without dripping or sagging. The material must have a low viscosity during pumping and spraying and a relatively high viscosity when it forms a film coating on the substrate surface. This is made possible by partial evaporation of the solvent as the sprayed mass goes through the air before impinging on the substrate surface. The distance between the spray head and the substrate and the rate of application are two important controlling factors.

2.7 Spread and Entrapment of Gas in Liquid and Solid Masses

Dispersion of gas in the form of tiny cells or bubbles in a liquid or solid mass constitutes some other kinds of tricky technology.

If a gas is passed through a liquid, gas bubbles will rise to the surface, break there and escape. For a more viscous liquid, bubble rise and gas escape will take a longer time. Using an emulsifier or surface active agent in the liquid may allow the bubbles to persist, leading to the formation of a froth on the liquid surface. The presence of a surface active agent not only reduces surface tension but also increases surface viscosity, thereby helping bubble formation and bubble stability and leading to foaming.

The low surface tension effected by the incorporation of a surface active agent allows the bubble to form readily and, if no other hindering forces are brought into play, the bubble soon expands, its walls become thinner and the bubble will eventually rupture. However, a higher viscosity at the gas-liquid interface, i.e. at the bubble wall, will confer stability to the foam. A medium viscosity of an optimum degree often plays a more critical role in foam stabilization than the lowering of surface tension by a surface-active agent.

All the techniques employed for making plastic- or rubber-based sponges or foamed/expanded objects require a gas to be released and entrapped within the matrix material. From the structural viewpoint, two basic types of foam are recognized, viz. the open-cell type and the closed-cell type. In closed-cell foams, the cells or voids are separate, discrete bubbles all through the mass, while the open-cell type has interconnecting void channels between the cells. Making a foamed object from rubber latex (mattresses, cushions, etc.) is initially a case of gas (air) in liquid (rubber latex formulation) dispersion, which finally, on gas entrapment in the cross-linked solid rubber matrix, turns into a gas-in-solid dispersion.

2.8 Temperature Dependence of Bulk Properties of Polymers

For practical purposes, all linear, amorphous polymers can behave as *Hookian elastic* (*glassy*) materials, *highly elastic* (*rubbery*) substances or *viscous melts* according to the prevailing temperature and time scale of experiments. Different property ranges exhibited by the same polymer at different temperatures are intrinsically related to variations in the physical structures or arrangements of the chain molecules as an outcome of different types and degrees of deformation.

2.8.1 Deformation Patterns and Transitions

A schematic diagram giving a plot of *log* (*shear*) *modulus* vs *temperature*, shown in Fig. 2.7, exhibits a number of *transitions*, each of which is connected with the gradual development of a specific type of molecular movement[1].

Below the respective transition temperature, the additional molecular mobility in question remains dormant or frozen in. With increase in temperature, and in the vicinity of and just beyond the transition point in question, the interplay of an additional molecular movement begins and grows to prominence to contribute to the deformation mechanism; the net result is a fast and prominent lowering in the resistance to deformation, i.e. modulus.

2.8.2 Glassy State and Hookian (Elastic) Deformation

In one extreme, at very low temperatures, the prevalent state of existence for a polymer is known to be the *glassy state* and the only deformation that comes

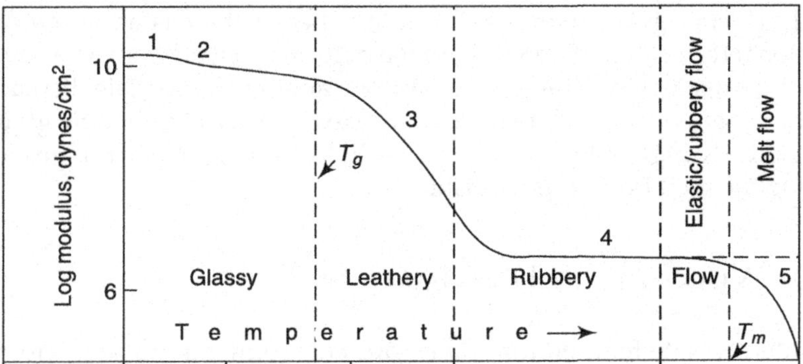

Defor-mation	1 Hookian glassy	2 Secondary transition	3 Primary transition	4 Highly elastic rubbery	5 Flow
Molecular motion	Bend & stretch of bonds	Excitation of internal rotation			
		Side groups	Main chain, gradual	Main chain (segmental) large-scale mobility	Chain slippage

FIGURE 2.7 Deformation patterns and molecular mechanism of deformation for a high polymer at different temperatures: plot of log (shear) modulus vs temperature

into play is Hookian elastic deformation, which is time independent and mechanically reversible. The deformation of a low degree occurring under an applied stress takes place instantaneously and the deformation also disappears instantaneously and completely with the recovery and restoration of the initial dimension on removing the applied stress. The deformation is not only mechanically reversible, but also thermodynamically reversible in view of the fact that no energy is dissipated during the stress-strain cycle (upto the elastic limit, as in Fig. 2.1). The Hookian deformation is attributed to the small changes in valence distances and valence angles (bend and stretch of bonds) which is manifested by the action of the applied stress against the strong binding forces of the homopolar bonds (dissociation energy 30–200 kcal/mol); therefore, it is associated with a high modulus, in the region of 10^{10} dynes/cm^2 or higher[1–3].

2.9 First Order Transition (Melting)

In the other extreme, at temperatures higher than the melting temperature (T_m) or the first order transition temperature, flow, which is completely irreversible, both mechanically and thermodynamically, takes place mainly due to the manifestation of *chain slippage*. The deformation energy is dissipated as heat in the viscous or viscoelastic flowing mass or melt and little configurational free energy is left after deformation; hence, there is no recovery or dimensional restoration. Even though the deformation mechanism at this stage is the slippage of chain molecules past one another, large scale displacement of chain segments normal to their chain axes is restricted, thereby giving rise to a relatively high melt viscosity. Cross-linked polymers are not prone to exhibiting the melt flow characteristics.

2.10 VISCOELASTIC DEFORMATIONS

The intermediate temperature range presents some interesting deformation behaviours. The prominent changes featuring a transition from a *glassy* region to a *rubbery* region via an intermediate leathery region, as in Fig. 2.7, are manifestations of *viscoelastic deformations* which are variously *time dependent* in nature, reaching an equilibrium at a given point after a certain time and recovering after about the same time at the specified temperature. The deformation for most polymers in the intermediate temperature region is mechanically reversible, more so if it is appropriately cross linked, but thermodynamically irreversible owing to time-dependency. A part of the deformation energy is stored as free energy that takes care of mechanical reversibility (dissipated as kinetic energy) and the remaining part of the deformation energy is dissipated as heat[1].

The transitions marked as *primary* and *secondary* may appear somewhat closer or separated depending on the nature, molecular weight and molecular weight distribution of the polymer[1]. The secondary transition arises only in limited cases as a result of inducing thermal excitations involving the rotation of the side groups of the polymeric chains; it is associated with only a small lowering in modulus as the temperature is raised. The inception of the limited rotation of segments of the main chains may also be attributed to such transitions, even though such segmental rotation about bonds is still strongly hindered and the chain molecules as a whole remain by and large strong and rigid.

2.11 GLASS TRANSITION OR SECOND ORDER TRANSITION

The primary transition occurring at a temperature commonly called the *glass transition temperature* (T_g) arises as a consequence of the notable excitation caused by the rotation of the main chain segments. Almost all polymers are characterized by a primary transition or a *glass transition* but many of them may not be associated with a distinct secondary transition. Through the glass transition, the polymers undergo a sharp, large change in modulus. Below the glass transition temperature (T_g), the polymer is in the glassy state, showing a prominent, dimensionally stable, rigid and often brittle character and high strength and modulus, while above T_g the polymer assumes a *leathery to rubbery* character depending on how farther away it is from T_g. In the temperature region just above or close to T_g the polymer is more leathery, while at a somewhat higher temperature and further away from T_g it turns rubbery and continues to be so over a long *rubbery plateau* region, showing high deformations under rather low stresses and, if cross-linked appropriately, further showing notable toughness and shock absorbing properties as a corollary to the exhibition of long range elasticity, i.e. high instant elongation/deformation on stress application and instant recovery to original dimensions without permanent set on stress release.

Almost all polymers are characterized by a glass transition temperature (T_g), which is different for different polymers[1]. Below the corresponding glass transition temperature, a polymer is dimensionally stable, hard, rigid, strong and often brittle, while above T_g it is usually soft, readily extensible, rubbery and tough. Values of T_g and T_m for some common polymers are shown in Table 2.1. A polymer having its T_g far below room temperature is conventionally viewed as a rubber and polymers having T_g far above the room temperature appear glassy and are viewed as plastics. Polymers whose glass transition temperatures are close to room temperature are useful neither as rubbers nor as plastics. The glass transition temperature T_g can be lowered by copolymerization (internal plasticization) or by plasticizer incorporation (external plasticization); T_g can be enhanced by cross linking and fibre reinforcements. Glass transition temperature is also referred to as the *second order transition temperature*[1].

TABLE 2.1 Melting temperature (T_m) and glass transition temperature (T_g) of some polymers

Polymer	*Repeat Unit*	*Tm, °C*	*Tg, °C*
Polyethylene	$—CH_2—CH_2—$	137	–115, –60
Polyoxymethylene	$—CH_2—O—$	181	–85, –50
Polypropylene (isotactic)	$—CH_2—CH(CH_3)—$	176	–20
Polyisobutylene	$—CH_2—C(CH_3)_2—$	44	–73
Polybutadine (1, 4 cis)	$—CH_2—CH{=}CH—CH_2—$	2	–108
Polyisoprene (1, 4 cis), (NR)	$—CH_2—C(CH_3){=}CH—CH_2—$	14	–73
Poly (dimethyl siloxane)	$—OSi\,(CH_3)_2—$	–85	–123
Poly (vinyl acetate)	$—CH_2—CH\,(OCOCH_3)—$	—	28
Poly (vinyl chloride)	$—CH_2—CH\,Cl—$	212	81
Polystyrene	$—CH_2—CH\,(C_6H_5)—$	240	95
Poly (methyl methacrylate)	$—CH_2—C(CH_3)\,(COOCH_3)—$	200	105
Poly tetrafluoroethylene	$—CF_2—CF_2—$	327	126
Poly (caprolactam)	$—(CH_2)_5\,CONH—$	215, 223	50
Poly (hexamethylene adipamide)	$—HN(CH_2)_6—NHCO—(CH_2)_4CO—$	264	53
Poly (ethylene terephthalate)	$—O(CH_2)_2—OCO—(C_6H_4)\,CO—$	254	69
Poly (ethylene adipate)	$—O(CH_2)_2—OCO—(CH_2)_4\,CO—$	50	–70

The full loading-unloading cycle in most rubbers is characterized by a phenomenon called *hysterisis*; the *unloading curve* in the stress-strain plot is seldom a retrace of the *loading curve*, but is usually one that passes below the loading curve. The loop thus formed between the loading-unloading curves as in Fig. 2.8 is known as the *hysteresis loop*. The area of the hysteresis loop gives a measure of the heat evolved over the loading-unloading cycle. In most dynamic applications, the heat generated over successive cycles keeps accumulating, as the rubbers in common use are insulators. If the heat is not allowed to be duly dissipated by conditioning the environment, by intermittent stoppage or by phasing out the cyclic process, it would be difficult to avoid heat accumulation and a fatigue failure, as a consequence of the associated progressive, concomitant thermo-mechanical and thermo-oxidative degradations. The right dose of a selected antioxidant and antiozonant contributes stability against thermo-oxidative degradation.

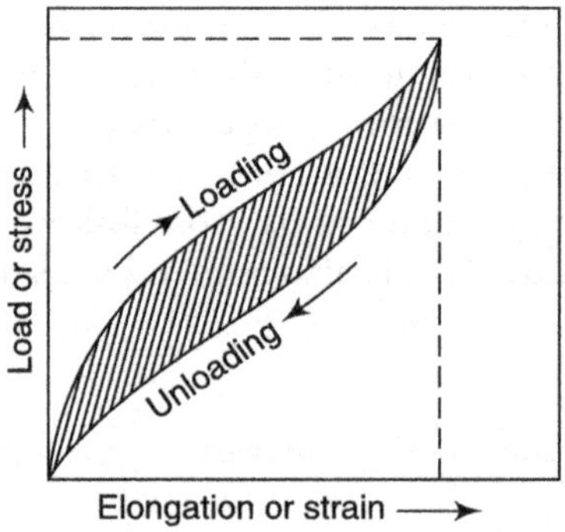

FIGURE 2.8 Stress vs strain curves over a loading-unloading cycle, showing a hysteresis loop

2.12 TEMPERATURE DEPENDENCE OF MELT-VISCOSITY

The molecules—or more importantly, the chain-segments of the molecules of a polymer melt or a liquid polymer under the prevalent thermal condition—normally remain confined to fixed mean positions under the influence of the forces exerted by the surrounding molecules or chain segments. However, on occasions, a chain segment adjacent to a vacant site, as depicted in Fig. 2.9, will overcome the influence of the neighbouring molecules or chain segments to jump or move into the molecular hole or segmental gap, thereby creating a new hole or vacant site in its previous position.

Thus, the structural pattern of chain segments and the relative positions of vacant sites undergo constant changes as a consequence of jumping molecules, chain units or segments into vacant holes. The changes would be random in the absence of any external/applied stress and little overall flow would result because of the lack of direction for jumping. Under a shear action, the random effect gives way to a directional effect and jumping to neighbouring vacant

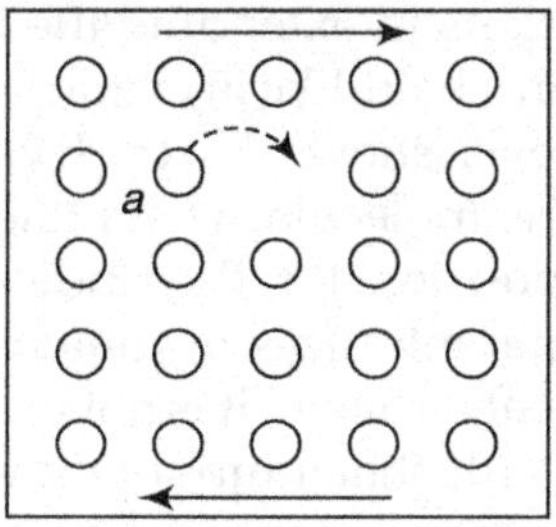

FIGURE 2.9 Model of a liquid as a mass of a regular array of molecules with occasional vacant sites, or molecular holes, as indicated by the position of the dotted arrow head

sites in the direction of the applied stress is manifested into a net movement or flow in the said direction.

Extending the molecular model described by Eyring[5] for melts of low molecular weight materials to polymer melts as well, flow ensues as a consequence of the independent jumping of individual molecules (for small molecules) or of small chain segments (for chain polymers), but not really due to the simultaneous sliding of all the chain molecules or chain segments in the layer. To break away from the binding or attractive forces of the neighbours, a jumping molecule or chain segment must acquire a certain amount of energy called *activation energy*. A higher temperature raises the probability of a given molecule/chain segment acquiring this energy. An Arrhenius-type equation relating the viscosity (η) to the jumping frequency (A) and the absolute temperature (T) may then be expressed, see equation (2.5), where E is the activation energy and R, the universal gas constant. Figure 2.10 gives an idea of the temperature dependence of the melt viscosities of poly (methyl methacrylate), PMMA, polystyrene, polyethylene and ethyl cellulose; the presented data are much in tune with equation (2.5):

$$\eta = A\, exp\, (E/RT) \tag{2.5}$$

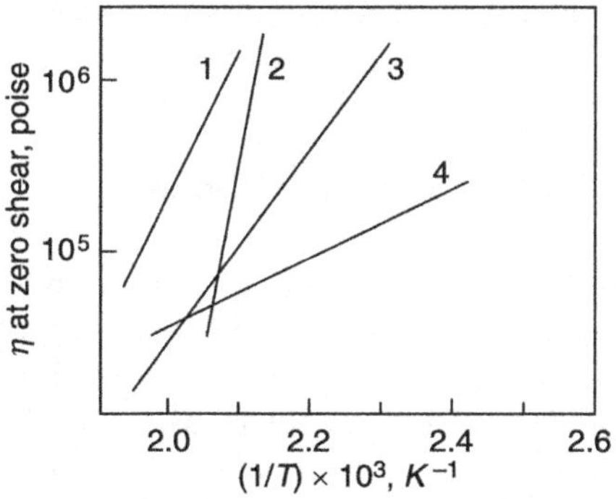

FIGURE 2.10 Effect of temperature (T) on melt viscosity (η) of some polymers: (1) PMMA, (2) ethyl cellulose (3) polystyrene and (4) polyethylene

There may be some reservations in extending the Eyring theory to polymers since polymer molecules are chain-like in character and not really spherical. Experimental analysis, however, shows that the E values for polymer melts are not much higher than for low molecular weight liquids/melts of comparable chemical nature and composition. It is thus indicated that the flow units in the two systems are of comparable sizes or dimensions. In reality, as has already been stressed for a polymer system, it is not the whole molecule, but a short segment of the chain molecule that jumps or moves at a given instant. A specific whole chain can seldom move, jump or slip past as a single entity, but separate short chain segments move practically independently of each other.

Further, the temperature dependence of melt–viscosity and the activation energy obtained from a study like this are mostly independent of chain length, but the melt–viscosity under a specified condition as such is largely dependent on chain length because the viscosity value is related to the mobility of the chain molecule as a whole.

2.13 Molecular Weight Dependence of Melt-viscosity

This feature is schematically shown in Fig. 2.11. The plot is characterized by a sharp change at a particular molecular weight value, say ≥ 20,000. The slopes of the two linear portions of the plot below and above this critical molecular weight M_c are about 1.75 (over the lower molecular weight region) and 3.4 (over the relatively high molecular weight region). The exact value of M_c may vary for different polymers. The transition in the viscosity behaviour using zero shear or low shear rates points to additional hindrance to flow from this point onwards, consequent to notable enhancements in chain entanglements. Below this point, the chain molecules apparently move independently as in common low molecular weight liquids, but above this point, the entanglement of chain molecules become so high and prominent that the movement of one causes others to drag along with it; as a result, there is a much sharper viscosity rise with further rise in molecular weight [6–9].

2.14 Gough–Joule Effect for Rubbers

Gough[10] (1805) and Joule[11] (1859) observed the phenomenon of reversible contraction on heating stretched rubber and the phenomenon is commonly known as the Gough–Joule effect. An elegant demonstration of this effect can be simply given by the rubber-spoked wheel, as in Fig. 2.12.

A light metal (Al or Al alloy) rim (cut from a 3–4 mm sheet and formed into the rim) having a large number of equally spaced saw-cuts around its circumference is turned to a wheel. Equal length rubber bands stretched to about 300% of their original length are fitted around the cuts on the rim. The wheel of nearly 25 cm diameter is mounted on a standard frame and suspended

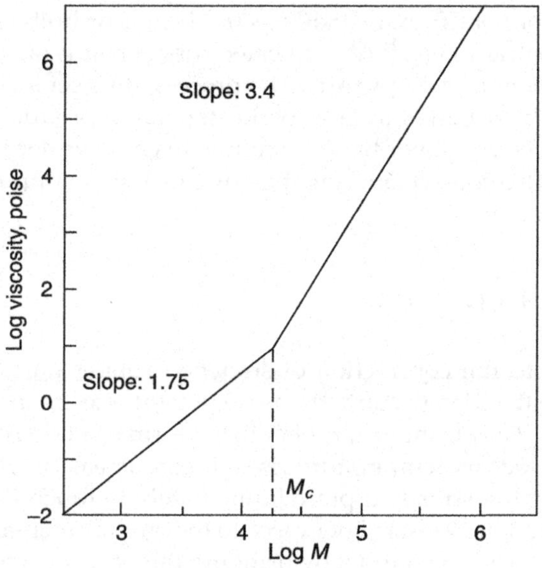

FIGURE 2.11 Dependence of polymer melt viscosity, on molecular weight (M): Plot of log viscosity (poise) vs log M

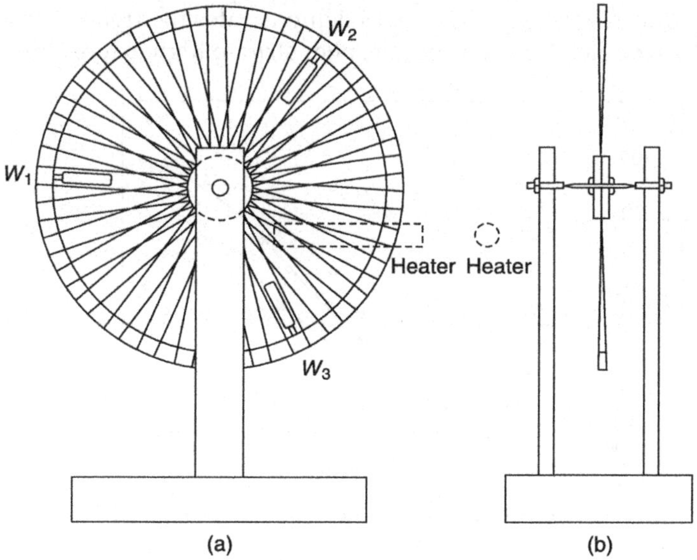

FIGURE 2.12 Rubber-spoked wheel for demonstration of the Gough–Joule effect: mild heating of stretched rubber bands (forming the spokes of the wheel) causes the wheel to rotate continuously

freely by ball bearings to eliminate friction and facilitate rotation in a vertical plane. The wheel is initially imparted a fine balance by adjusting the small weights W_1, W_2 and W_3 by working on screws attached to the rim.

On application of heat from a small electric heater or bulb (100 W) from one side of the wheel, as in Fig. 2.12, the heated rubber bands facing the heater contract, making them move upwards. The process, thus set in, brings in the next few lots of rubber bands to face the heater, again producing the same result. As the successive rubber bands continue to come under the influence of the heater, they also contract in turn, thereby causing a continuous rotation of the rubber-spoked wheel.

2.15 Heat of Extension

Intimately related to the contraction of stretched rubber on heating is the development of heat and consequently, a rise in temperature in a rubber test piece on extension. If a specimen of rubber in the form of a thin film is quickly stretched or extended, its temperature rises; it can be easily felt or demonstrated by allowing the stretched piece immediately to touch the lips. After equilibrating the stretched test rubber piece to the environmental/room temperature, if it is allowed to contract by bringing the opposite ends closer together, it will be found to be noticeably cooler, without however, allowing the test piece to snap back by releasing one end, in which case the kinetic energy is dissipated and no cooling will occur. The development of heat in rubber on stretching is thus reversible. Figure 2.13 demonstrates the trend of the rise in temperature on stretching a specified rubber strip to high extensions.

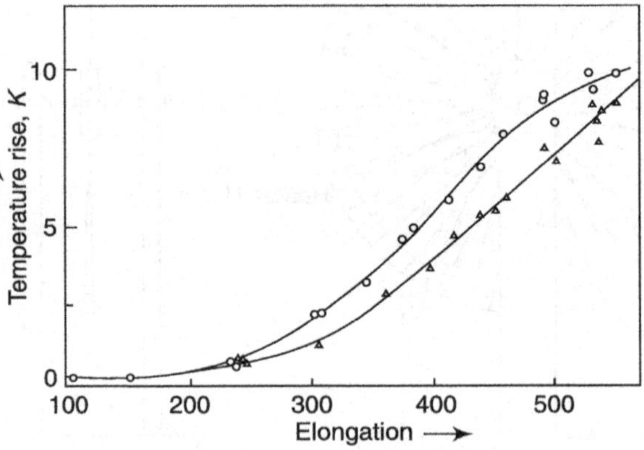

FIGURE 2.13 The trend of temperature rise in rubber on stretching it to high elongations

2.16 Creep and Relaxation Phenomena

Elastoviscous liquids or viscoelastic solids are liquids or solids respectively that show a mixed property range of truly elastic solids and truly viscous liquids. Their flow behaviour is time dependent, as mentioned earlier, and is characterized by *creep* and *relaxation* phenomena.

Creep or the cold flow of a polymer material, characterized by *viscoelasticity*, is the long-time, continuous deformation or strain that occurs under the influence of a low, constant load or stress. Creep or time-dependent strain under a low, constant stress may be either fully recoverable with time on stress release, or may involve permanent deformation with part recovery.

Relaxation, however, is a flow process that occurs in a viscoelastic material in which the stress required to keep the material strained to a constant amount of deformation decreases with time. The stress on the test strip may slowly drop to zero, as it suffers a permanent set to the extended length.

Metals and alloys and some ceramic materials undergo elastic deformation and can be strained to < 1% before they reach their elastic limits. Their deformation is a result of the distortion of normal bond lengths and bond angles. Polymeric solids, however, undergo much larger elastic deformations. Chain molecules of polymers remain in a highly coiled and folded form. If a specific coiled position corresponds to a state of minimum free energy, applying external tensile stress causes temporary uncoiling of the chain molecules, leading to a relatively large elongation of the test strip, followed by a return to the original dimensions as the stress is withdrawn. The retraction is considered to be aided by chain entanglements, thermal effects and limited cross linkages. Reversible elongation as great as 1000% may be exhibited by duly cured natural rubber and some of its synthetic counterparts. The long range or so-called kinetic elasticity of duly cured rubbers involves a change in the physical structure of the chain molecular aggregates, leading the randomly-coiled, amorphous mass of the unstretched rubber chains to a parallelized array of crystalline, fibrous mass of distinctly different property range and phase morphology, due to stretching them to very high (~ 1000%) elongations even on applying a relatively low stress.

2.17 Distinctive Strain Geometry of Rubbers

For (vulcanized) rubber, the strain developed is not proportional to the stress applied and a typical non-linear stress-strain (L/L_0) or force-extension plot is shown in Fig. 2.14. The deviation from Hooke's law is obvious and for rubbers, the deformations observed are very large indeed (> 600% without a break) as against upto 1%–2% for most engineering solid materials (steel, alloys, glass and ceramics), where Hooke's law is applicable. For very large elastic deformations, observable only in rubber-like materials, an altogether different approach is required to define and describe the elastic property, called *long range elasticity*, necessitating a concept of *molecular mechanism* related to *uncoiling* or *unfolding* of network chain segments producing large deformations without materially changing the volume. The forces required to produce such network deformations are comparatively weak. This is why the modulus of elasticity of rubber is so much lower than that of an elastic material such as steel.

The volume of a vulcanized rubber specimen is primarily determined by the actual volume occupied by the chain molecules themselves and is in no

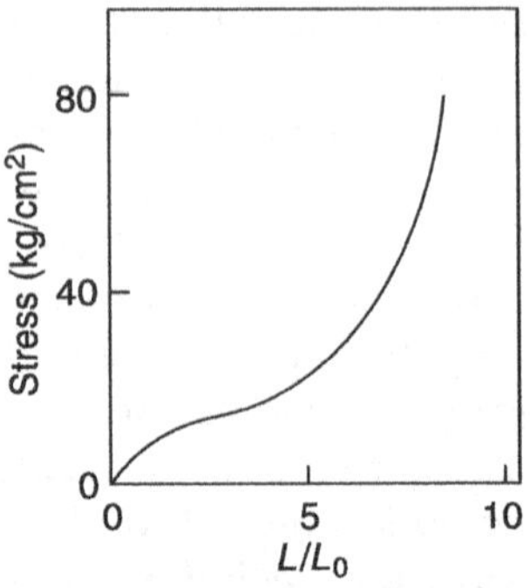

FIGURE 2.14 Stress-strain (L/L_o) plot for natural rubber gum vulcanizate: for the test strip, L_o = initial length and L = observed instantaneous length on specified stress application

way related to chain conformations or the existence of a cross-linked network structure. In fact, the vulcanization or curing process itself brings little change to the density of rubber. The volume or density of the material is merely a function of *molar cohesion*, just as for any other material. The large deformation of the rubber network is quite a separate process, linked to chain unfolding, that readily takes place without affecting the lateral forces between chains or chain segments—without affecting the volume and hence the density.

However, in an elastic material such as steel or glass, the very small deformation on stress application is primarily due to the bend and stretch of bonds in the molecules; the ratio of lateral contraction to the longitudinal extension, called Poisson's ratio, has a particular value in each such system, and there is an increase in volume on extension consequent to stress application; the amount of this increase, directly related to Poisson's ratio varies from one material to another.

2.18 Williams–Landel–Ferry (WLF) Equation

In polymer systems, the glass transition corresponds to the onset of long-range, liquid-like mobility of molecular segments characterizing the rubbery state. The enhanced mobility at or above T_g leads to the utilization of more space or volume than what is needed and is used for short-range movement of chain-bound atoms or groups in the glassy state. The increase in the relative free volume with the rise in temperature (T) above the glass transition temperature (T_g) leads to a much enhanced volume expansion coefficient in this region. The fraction f of the 'free' volume may be expressed as

(i) $f = f_g$ for $T < T_g$ and

(ii) $f = f_g + (T - T_g)\,\Delta\alpha$ for $T \geq T_g$

Thus, f is a constant at its minimum value for all temperatures below T_g, which thereby becomes a reference point; for $T > T_g$, additional free volume created by the thermal effect becomes dependent on the increase in the expansion coefficient ($\Delta\alpha$), such that we may write:

$$f = f_g + (T - T_g)\,\Delta\alpha \text{ for } T \geq T_g.$$

The WLF equation relates the log of relative viscosity (η/η_g) with the temperature difference ($T - T_g$) by an expression as shown in equation (2.6):

$$\log \frac{\eta}{\eta_g} = \frac{-\alpha(T - T_g)}{b + (T - T_g)} \tag{2.6}$$

Analysis further points to the fact that the ratio η/η_g also gives the ratio of the rates of Brownian motion of the molecular segments at T and T_g. Further, assessment of available data on the η-T relationship and their interdependence for many glass-forming polymers fixes the values of the numerical constants a and b in the WLF equation as 17.4 and 51.6, respectively. It is thus implied that, the viscosity of the polymer should become infinite at $T = (T_g - b) = (T_g - 51.6)$, indicating thereby that for polymers, all molecular motions should get totally frozen at $T < (T_g - 51.6)$. The point of actual importance, however, is that an altogether new mechanism of deformation takes over or becomes overwhelmingly prevalent as the test temperature T falls below or much below T_g.

Many elastomer adhesive joints fail cohesively within the adhesive layer when tested at low rates and high temperatures, but they show clean interfacial separation when tested at high rates and low temperatures. This is a consequence of the viscoelastic nature of the adhesive material. Polymeric adhesives become liquid-like and weak at elevated temperatures when their cohesive strength becomes less than that of the interface. So, in this case, failure occurs within the adhesive layer. However, with lowering of temperature, the cohesive strength of the adhesive layer increases. As the cohesive strength surpasses the interfacial strength, the locus of least resistance becomes the interface.

References

1. Ghosh, P, *Polymer Science and Technology—Plastics, Rubbers, Blends and Composites*, 2/E, Tata McGraw-Hill, New Delhi, 2002, Ch. 7 and 10.
2. Schmidt, A X and C A Marlies, *Principles of High Polymer Theory and Practice*, McGraw-Hill, New York, 1948.
3. Severs, E T, *Rheology of Polymers*, Reinhold, New York, 1962, and Tordella, J P, *Rheology—Theory and Applications*, Vol. 5, F.R. Eirich, ed., Academic Press, New York, 1969, Ch. 2.
4. Powell, S M et al, *Modern Plastics*, Vol. 28, 1951, 129.
5. Brandrup, J and E H Immergut, eds, *Polymer Hand Book*, Interscience, New York, 1967.
6. Eyring, H, *J. Chem. Phys.*, Vol. 4, 1936, 283.
7. Flory, P J, *J. Amer. Chem. Soc.*, Vol. 62, 1940, 3113.
8. Fox, TG and PJ Flory, *J. Phys. Chem.*, Vol. 55, 1951, 221.

9. Fox, T G, S Gratch and S Loshack, *Rheology—Theory and Applications,* F R Eirich ed, Academic Press, New York, 1956, Vol. 1, Ch. 12.
10. Gough, J, *Proc. Lit. and Phil. Soc., Manchester, 2nd Ser.,* Vol. 1, 1905, 288.
11. Joule, J P, *Trans. Roy. Soc.,* A 149, 1859, 91.
12. Treloer, L R G, *Introduction to Polymer Science,* Wykeham, London 1970, Ch. 3.

Chapter

3

Glues, Resins and Polymers Used in Adhesives and Coatings

The science and technology of adhesives and coatings are deeply rooted in the preparative science and technology of the film-forming glues, resins and polymers in powder, melt, solution and dispersion forms; manipulations of their spreading and film-forming processes through appropriate articulation and control of solid or volatile content, stability, rheology, tack, adhesion to surfaces and setting or drying of the applied formulations to the right film thickness on the surface of substrates are also important considerations.

The early developments of adhesives and (surface) coatings applications were more empirical than based on scientific cause-effect analysis and understanding. The development of process technology preceded the development of a scientific understanding of material selection and the analysis of the physico-chemical principles involved. An increasingly complex spread and combination of glues, gums, resins, high polymers and many special additives are being used as adhesives and (surface) coatings with growing emphasis in their efficient manufacture and effective formulations.

3.1 Base Materials for Natural Glues

3.1.1 Protein Glues

Among the amide and amine type glues or adhesives, some are developed and made available as water based, natural glues. These adhesives derived from collagen and other proteins are losing importance now, and making room for grades based on starch/dextrin; more popular than even these are the synthetic water-based adhesives derived from poly (vinyl acetate) and poly (vinyl alcohol). The reason for the decline in protein (animal and fish) glues derived from collagen is cost.

Likewise, another protein-based adhesive, derived from casein and once widely used, has almost disappeared due to the non-availability or high cost of skimmed milk—indigenous or imported.

Blood-glues, consisting of serum albumin and globulin, have also been losing importance. They were once in demand as plywood adhesives and for veneering. Now-a-days, blood glues have almost completely been replaced by phenolic adhesives. Beef and hog blood have so far been used as the main sources of blood glues.

The fibrous protein collagen that occurs universally in connective tissue and skin has been one of the widely available and studied proteins[1,2]; its structure involves three intertwined helices of the type shown in Fig. 3.1. Collagen is a very unusual protein in terms of composition. Nearly or over 25% of its residues are proline or hydroxy proline derived from the corresponding cyclic amino acids. The other major amino acid residue of collagen (24%) is glycine. Three levels of complexity exist in a given protein structure. The primary structure relates to the specific sequence of a set of amino acids in the polypeptide chain; the secondary structure relates to the way the protein molecular chain is coiled, perhaps to form an α-helix; the tertiary structure is the way the coiled chain is folded and hydrated in the natural state.

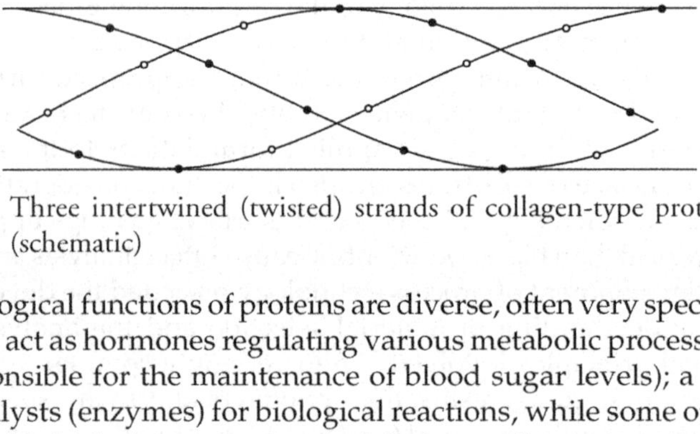

FIGURE 3.1 Three intertwined (twisted) strands of collagen-type protein molecules (schematic)

The biological functions of proteins are diverse, often very specific and complex. Some act as hormones regulating various metabolic processes (e.g. insulin is responsible for the maintenance of blood sugar levels); a set of others acts as catalysts (enzymes) for biological reactions, while some others are recognized as biological structural materials (e.g. collagen in connective tissue and skin, and keratin in hair). The oxygen-carrying properties of haemoglobin in mammals and other animals, birds, etc. are well known. Some blood proteins form antibodies, which provide resistance to disease, while nucleoproteins are important constituents of the genes that transmit and supply genetic information, the core function in cell biology.

Collagen molecules are long and thin and each appears to be made up of three twisted polypeptide strands[2] as shown schematically in Fig. 3.1. When collagen is boiled with water, the strands come apart; the product is ordinary cooking gelatine. Collagens from connective tissue and skin are made up of fibrils (200–1000 Å wide) as revealed by X-ray diffraction photographs. The fibrils are actually bunches of collagen molecules running parallel to the long axis. Electron micrographs reveal regular bands nearly 700 Å apart across the fibrils and the collagen molecules virtually appear aligned in the fibrillar direction[2] but regularly staggered by nearly a fourth of their length, as schematically shown in Fig. 3.2.

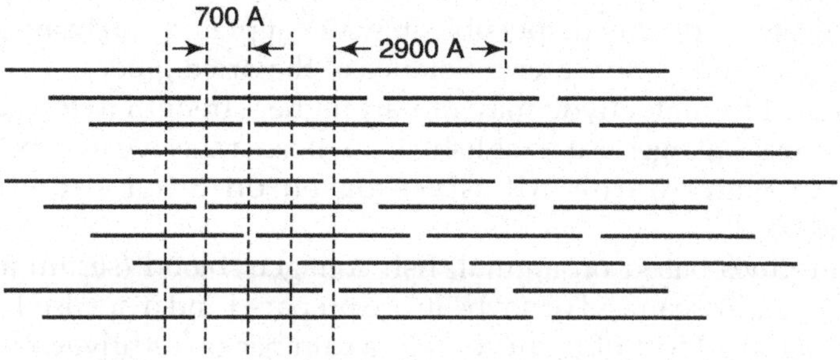

FIGURE 3.2 Arrangement of collagen molecules in a fibril with 700Å spacing as traceable in an electron micrograph (schematic)

Casein, a protein from skimmed milk is heated to 40–50 °C and coagulated as a curd by controlled addition of acid. The coagulant (curd) is separated from the whey, washed to remove acid and salt and most of the water is squeezed out. The curd is then mixed with an appropriate dose of a preservative, dried and packed for delivery to glue-making or fibre-forming manufacturing units.

Blending different batches of raw materials is an art practiced in these manufacturing units. After blending, the casein mix is dispersed in aqueous caustic soda, producing a sticky, gelatinized solution. Nearly 2.5%–2.7% of caustic soda of the weight of casein is used. Standard preservatives in the right doses ensures safeguard against instability from insect/microorganism attack during storage and application.

Seaweed, usually *Laminariae*, also serves as a source of glue and fibre. The weed is collected, dried and milled to powder. It is safe to store it in a dry form; otherwise it would quickly rot from microorganism attack. The powdered weed is treated with a solution of sodium carbonate and caustic soda to convert all the alginate in the weed to Na-alginate. The clear, viscous Na-alginate solution is ready for use as glue when mixed with a small dose of preservative. The insoluble/suspended impurities are centrifuged and the solution is then bleached and sterilized, if necessary by the addition of sodium hypochlorite. The alginic acid can be isolated from this extract by acidification, which may then be purified and dried. The alkali solution of the purified alginic acid is straight away used as glue or for spinning to an acid/salt bath to get alginate fibres. Alginate and agar gels or glues duly filled with calcium hydroxide, zinc oxide, fine silica, etc. and appropriately swollen with water to a desired plasticity are useful as dental (gum and tooth) impression and duplicating materials for dentistry.

Soybean (35% protein content) is also viewed as an attractive source of protein. The beans are dried and crushed and their fat is extracted with n-hexane. The residue is treated with a dilute solution of sodium sulfite, whereby the protein part gets dissolved in it. The protein gets reprecipitated on acidification to pH 4.5. It is then dissolved in NaOH solution, and may be directly used as glue or spun into an acid/salt bath for making soybean protein fibres. Fat-free whole soybean meal (flour) actually a mixture of proteins and carbohydrates is readily dispersible in water at pH $\geq$ 11 on heating. Lime may be used to enhance the water resistance of the dried glue.

Small doses of formaldehyde may impart higher strength to the adhesive line because cross-linkages are established. Soybean adhesive may be readily blended with water-borne adhesives based on other proteins and polysaccharides.

All protein glues based on animal/fish collagen, blood (serum albumin and globulin), casein, seaweed or soybean are prepared and dispensed in aqueous, alkaline media. Most of them contain a number of additives, including fillers, stabilizers, some salts, formaldehyde, etc. that improve their consistency, spreadability binding quality and strength.

3.1.2 Starch and Dextrin Glues

Starch- and dextrin-based adhesives are almost exclusively water-based and are widely used for bonding paper and cardboard. They are also popular as bookbinders' adhesives. Starch used as adhesives on a large scale is virtually dextrin-based. Dextrins are products of starch hydrolysis under various conditions. White dextrins are mostly obtained by dry heating starch with acid. Higher temperatures of heating using low acid content yield light-coloured dextrins with ready solubility in water. Modifiers used in starch/dextrin glues are urea, thiocyanate, iodides and sodium nitrate. They improve the solubility of the adhesive base polymer, lower its viscosity and prevent or delay the precipitation causing development or enhancement of inhomogeneity. Borax ($Na_2B_4O_7$) helps in raising viscosity and adhesive bond strength. Urea-formaldehyde resin (10–20% by weight) helps in improving the water-resistance of the glue-line and further helps prevent delamination. All protein glues and polysaccharide-based adhesives are considered as bio-adhesives.

Starch/dextrin adhesives presently face stiff competition from synthetic vinyl adhesives like poly (vinyl acetate) and poly (vinyl alcohol). Gum arabic and agar were once popular in making postage stamp and relevant glues that were readily remoistenable for easy, effective pasting.

The fibrous tissue in the cell walls of plants contains the most abundant polysaccharide, *cellulose*—a molecule of which consists of long linear chains of glucose units linked to each other by *1, 4 β-glucopyranose* as in Fig. 3.3 (a). The second most widely distributed polysaccharide is starch, which is stored in the seeds, roots and fibres of plants as a food reserve and looked upon as a potential source of glucose. Molecules of starch are essentially polymers of glucose, linked to each other by *1, 4 α-glucopyranose* links, as in Fig. 3.3 (b). "*A "cellobiose"* unit is the repeat unit of cellulose chains while a *"maltose"* unit is the repeat unit of starch chains. Usually, in a given source of starch, there are two structurally different polysaccharides[2], each consisting of entirely

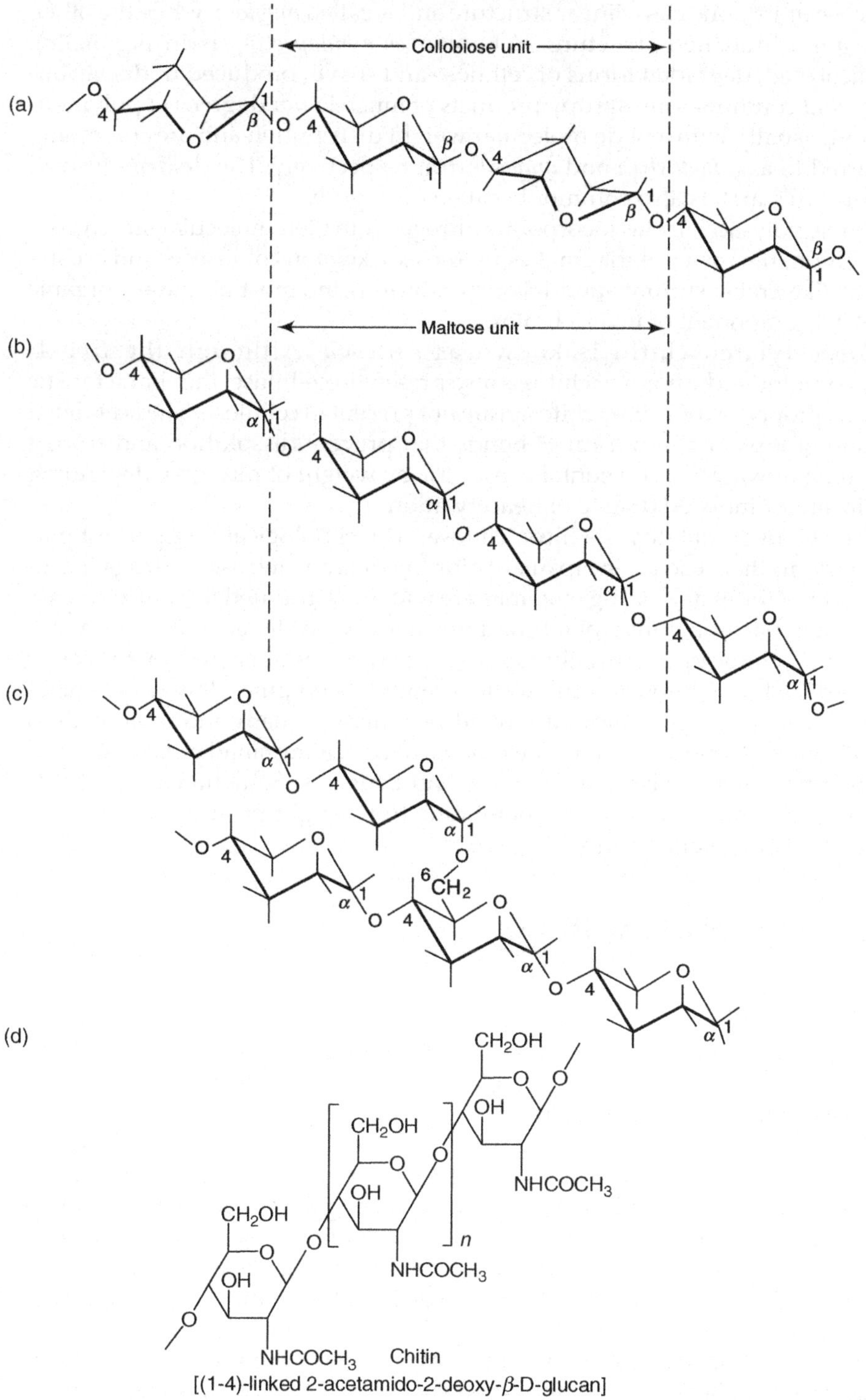

FIGURE 3.3 Structural segments of (a) cellulose, (b) amylose, (c) amylopectin and (d) chitin

glucose units : one has a linear structure and is called *amylose,* while the other, bearing a branched structure, is known as *amylopectin,* as in Fig. 3.3(c). Hydrolyzed, degraded forms of cellulose and starch, produced under steaming/acid conditions are syrupy products of much lower degrees of polymerization, usually with a wide molecular weight distribution and are commonly referred to as *cellodextrin* and *amylodextrin* respectively. The dextrin derived from corn starch is also commonly called *corn syrup.*

Some polysaccharides incorporate nitrogen into their molecular structure—e.g. chitin, the material that makes up the exoskeletons of insects and crustaceans like crabs, shrimps and lobsters. Chitin is the most abundant organic skeletal component of invertebrates.

Deacetylated chitin is known as *chitosan.* Although the β(1–4) anhydroglucosidic bond of chitin is also present in cellulose, the characteristic (basic) properties of chitin/chitosan are not present in cellulose. The acetylated amino groups of chitin form H-bonds that prevent dissolution and restrict swelling in water. Chitin contains 5%–8% by weight of nitrogen, depending on in situ or induced degree of deacetylation.

The biodegradability of chitin/chitosan is technologically important particularly in their use as encapsulants for the delayed release of drugs, pesticides, etc. Chitin degrading enzymes are widely distributed in nature and are present in bacteria, fungi, plants and the digestive systems of most animals.

Chitin/chitosan is virtually non-toxic, and is administered to accelerate wound healing. Chitosan is an excellent candidate for guided tissue regeneration in dental surgery, dermal wound healing and many more biomedical applications. Chitosan membranes are used for healing small or large injuries or wounds such as those from burns and bedsores. Sometimes absorbable sutures are used to attach the membrane to the right position due to non-availability of a suitable bio-adhesive.

3.2 Some Natural Resins

3.2.1 Shellac

An elementary analysis of *shellac* indicates an empirical formula, C_4H_6O; with a mean molecular weight of 1000. One believes that the formula $C_{60}\ H_{90}\ O_{15}$ represents its average molecule.

Shellac is the term used for the refined form of lac. Lac, in fact, is the resinous secretion of the lac insects (*Laccifer Lacca (Kerr)*), cultivated as parasites on certain trees in Madhya Pradesh, Bihar, West Bengal and some other states in India, and also in Myanmar (Burma) and Thailand. Most of the resin is secreted by the female insect. The trees on which the lac insect is cultivated in India are *Palas, plum* or *ber* and *kusum.* Lac is collected by cutting off the encrusted branches, which are then scraped to remove the lac, or chopped into small pieces that are commonly called *sticklac.* Next, the sticklac is mechanically crushed; the remains of the wood sticks are sieved off; it is washed with water to remove the lac dye, at least in part, and then dried and winnowed (for removing fragments of the sticks). Mechanized processes have been developed and the semi-refined product is now called *seedlac*. Further purification, through adoption of the age-old heat (melt) process, allows the molten seedlac to be squeezed through a cloth bag—the product obtained is called *shellac* and may be obtained by a hand-made or machine-made process. The solvent refining process (using industrial alcohol) ensuring efficient solvent recovery was later developed.

The resinous component of lac is a solid solution of the condensation products of some specified monobasic and dibasic hydroxy acids[3,4]. A major (30%–40%) constituent of shellac is *9, 10, 16 trihydroxy palmitic acid*, commonly called *aleuritic acid* (M.P. 100–101°C). Another known constituent (6–7%) is a dihydroxy dicarboxylic acid, as in Fig. 3.4, called *shellolic acid* (M.P. 200–201°C).

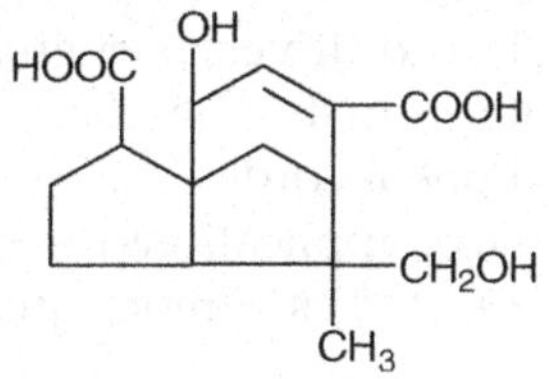

FIGURE 3.4 Structure of shellolic acid

The following chemical data are available for shellac.

TABLE 3.1 Some chemical properties of shellac

Acid value	65–75
Saponification value	225–250
Ester value	150–180
Hydroxyl value	260–280
Iodine value	10–20
Thiocyanogen number	18–20

Shellac can be split into two major fractions by ether. The ether-soluble portion, about 30% and called the "soft resin," is sticky and viscous. The ether-insoluble portion, called the "hard resin" or the "pure resin," has a melting temperature (94°C) higher than that of the whole shellac (80°C).

Shellac is a hard, somewhat tough, amorphous resin containing a small amount of natural wax and a yellow or red-yellow colouring matter. It develops enhanced brittleness on long standing and on heating for some time. It is insoluble in water, but soluble in alkaline, aqueous media (ammonia, amines, ethanol amines, soda, etc.). It readily dissolves in alcohols and the most commonly employed solvent is methylated spirit. The specific gravity of shellac is 1.20. It shows a tensile strength of 130 kg cm^{-2} and an elastic modulus of (13–26) $\times 10^3$ kg. cm^{-2}. The applications of shellac are based on one or more of its properties as given below:

(i) Shellac produces smooth, decorative and durable films that dry rapidly from its alcoholic (spirit) solution. These films exhibit good to excellent adhesion to a wide range of surfaces, giving high gloss, adhesion, hardness, strength and resilience. Shellac surface coating films are also resistant to sunlight and UV light
(ii) Alkaline (ammoniacal) solutions of shellac provide spreadable films of high adhesive power
(iii) Shellac is non-toxic and resists the action of hydrocarbon solvents such as petroleum products
(iv) Shellac is an excellent bonding resin with good to excellent spreadability to thin films from alcoholic solutions; it shows low thermal conductivity and a very low coefficient of thermal expansion
(v) It has a characteristically high dielectric strength, a low dielectric constant and is remarkable for its freedom from tracking
(vi) Shellac's ready fusibility and good thermal plasticity make it amenable to compounding with a host of additives, particularly fillers, and render it valuable for the moulding industry. Shellac was once (before 1960) popular in making gramophone records

Based on these properties, the major applications developed for shellac are[3,4]:

(i) Spirit varnishes—furniture polishes (French polish), undercoats and enamels
(ii) Glazes for pharmaceutical tablets and confectionary
(iii) Flexographic and other shellac-based printing inks
(iv) Aqueous varnishes for paper, wood and leather
(v) Hair lacquers and cosmetics
(vi) Electrical insulation (moulded insulators), insulating varnishes, laminated paper
(vii) Grinding wheels, sealing wax
(viii) Adhesives, cements and sealants
(ix) Fabric stiffening and water proofing
(x) Making abrasion-resistant, relatively stiff rubber compounds (10–20% shellac)

It must, however, be noted that in some of these application areas, shellac faces stiff competition from some synthetic resins (phenolics, petroleum resins, coumarone-indene resins, ester gum from rosin, alkyds, etc.)

3.2.2 Rosin

The natural *gums* and *resins* that have been in use for centuries as surface coating constituents are not polymeric. They are rather relatively large non-polymeric molecules of irregular structure and poor molecular symmetry, so they fail to crystallize well.

The chief constituent of the well known resin, *rosin*, the gum of pine trees, is *abietic acid*, $C_{19}H_{29}COOH$, having the structure shown in Fig. 3.5, and a molecular weight of 282. Physically, the plant gums and resins are hard and glassy.

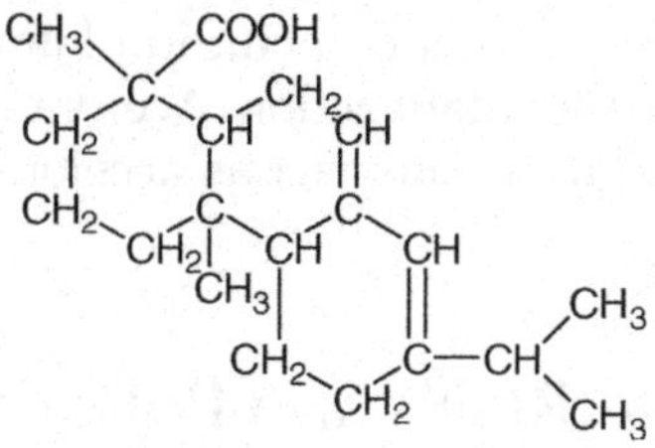

FIGURE 3.5 Structure of abietic acid

Common rosin (colophony) is the most plentiful resin produced in nature. Raw rosin is not considered as a good varnish (film forming) constituent. It can be readily spread into a thin film, which, however, develops minute cracks as the material crystallizes slowly. Its high acidity causes notable blooming that shows up as a bluish blush or discolouration of the coating, affecting transparency. Rosin, i.e. abietic acid, much like free fatty acids, reacts with basic pigments (ZnO, PbO, etc.) to form soaps, which may then adversely affect spreading and film formation, thus hindering the coating process. Raw rosin films may also develop undesirable tack under humid conditions, more so during the summer or monsoon.

Turning rosin (abietic acid) to its calcium salt, commonly referred to as *limed rosin* by cooking with lime, infuses some advantage. It is even better if rosin is esterified with glycerol in the absence of air, giving a product called *ester gum* of low acid value. The natural rosins, particularly their derivatives in the ester form are found to be more suited as additives or property-improving components in the preparation of polymers (say, alkyds) of specified surface coating property parameters.

3.3 NATURAL RUBBER (NR)

Natural Rubber (NR) is obtained as a milk-white latex from the rubber tree, *Hevea Brasiliensis* grown mostly in the equatorial climatic zones of South America, and South/South East Asia. The Hevea latex contains about 30–40% rubber hydrocarbon (*1, 4-cis polyisoprene*). The latex is collected in the rubber plantation field and small doses of ammonia (with or without sodium pentachlorophenate) are added to the latex as preservative simply to prolong its storage life. The rubber hydrocarbon is given a creaming treatment and then coagulated by adding acetic acid. Rubbers in bale forms contain 93–94% rubber hydrocarbon.

The coagulum contains nearly 80% water. Much of the serum is removed by washing and squeezing through hot/warm rollers. For *crepe* rubber manufacture, the wet strip is heavily milled on a roll mill and the sheets formed are finally dried and baled. For producing *smoked sheets,* the wet strips are lightly milled on the rollers and the ribbed sheets are finally dried in a smoke house. Spray-dried, "*whole latex*" rubber apparently contains all the solid impurities present in the rubber latex. Sulphur-accelerator systems are used as curatives.

The host of non-rubber constituents including latex proteins and (poly) phenols exhibit prominent antioxidant action. Acetone-extracted rubber ages rapidly and fatty acids present in latex act as accelerator activators during vulcanization.

3.4 Major Synthetic Resins and Polymers

3.4.1 Coumarone–Indene Resins

Coumarone and Indene (Fig. 3.6) are two major constituents obtainable from the fraction boiling at 165–180°C and derived from fractionation of coal tar naphtha.

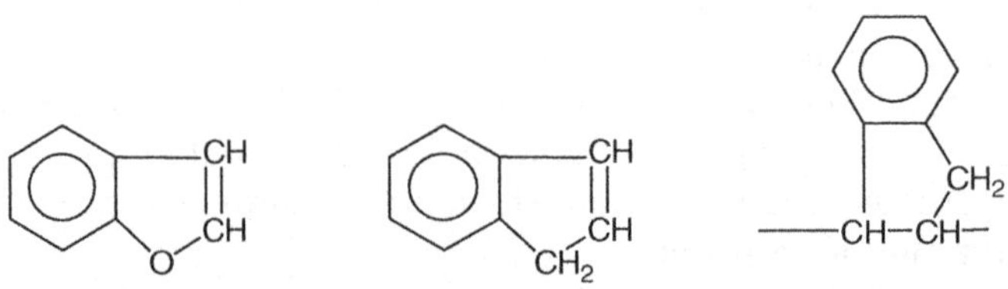

FIGURE 3.6 Structures of coumarone, indene and polyindene

Being aromatic in nature and having a vinylic structural unit attached to the aromatic ring, they closely resemble styrene. For resinification, they are polymerized in the crude naphtha without separation and isolation, using sulphuric acid as the (ionic) catalyst. The resin product is called *coumarone-indene resin*. The commercial polymers or resins so obtained are in the molecular weight range of 2000–2500, giving an average degree of polymerization (DP) of about 20–25. The resins are prepared in different grades, varying from sticky and soft to hard and brittle. They are variously used as process aids and property control agents in the formulation of *plastics, rubbers, adhesives, inks* and *coatings*. They are also useful as binders of a host of fibrous and particulate fillers, waxes, etc. for making hard floor tiles from vinyl thermoplastics.

Resins, likewise made from residues of some petroleum fractions and popularly known as petroleum resins also find use as binders and process aids in compounding selected plastics, rubbers, adhesives, inks and coatings.

The binding material in a given plastic or rubber product is invariably a polymer or resin, and the two words are often used interchangeably. However, a resin is commonly understood to be a material primarily organic in nature, either natural or synthetic in origin, which has the capacity to produce a thin, cohesive film either from its melt or its solution. A resin may be polymeric (characterized by high molecular weight and a characteristic repeat (chemical) unit in its structure) or non-polymeric (characterized by a relatively low molecular weight and having no identifiable repeat chemical unit in its structure). According to this resin concept, almost all polymers may be considered as resins, but not all resins are polymers. Conventionally, however, resins are brittle materials, showing different degrees of hardness under ambient conditions, even though some synthetic resins may appear as viscous liquids and as soft, tacky material at an intermediate stage of their growth.

3.4.2 Polyolefins

Polyolefins, viz. polyethylene (PE), polypropylene (PP), higher poly (α-olefins) and olefin copolymers are commercially produced by addition polymerization or chain growth polymerization of selected monomers or monomer combinations. Vinyl, divinyl or diolefin and related monomers are likewise polymerized by addition or chain growth mechanism.

(a) Low-density Polyethylene (LDPE)

For the polymerization of ethylene, a high pressure (1500–3000 atm), free radical polymerization process using traces of air or oxygen as catalyst was the early development; this process, developed in the mid to late 1930s in UK and USA, brought into the market the first generation polyethylene (PE) called *low density polyethylene (LDPE)* of the density range 0.92–0.945 g/cm^3. The reason for its density becoming lower than expected (0.96–0.98 g/cm^3) is the development of varied degrees of branching and even some sub-branching (introduction of 20–60 short and long branches per 1000 linear C-atoms in the main chain) consequent to intramolecular and limited intermolecular polymer (chain) transfer, thus causing significant deviation from the linearity of the chain molecules formed[5,6]. Branching caused a significant lowering in the melting point (110–117°C), the degree of crystallinity (60%–80%), hardness and rigidity (tensile strength 12–20 MPa, breaking elongation 200%–600%) and environmental stress cracking resistance for LDPE. It proved to be a useful, soft, flexible thermoplastic that showed an excellent heat-sealing property, particularly in thin film form, which was soon taken advantage of in the packaging field and in having extended heat-sealed films suited for canal lining, agriculture films for mulching and in many water proofing or water-repelling applications.

(b) High-density Polyethylene (HDPE)

The technology of making high density (0.945–0.965 g/cm^3) and predominantly linear polyethylene under much lower pressures was developed at a later stage in the mid 1950s with the development of the Ziegler-Natta catalysts based on aluminium trialkyl ($AlEt_3$) and transition metal halides such as $TiCl_4$ and $TiCl_3$. Almost simultaneously, some supported metal oxide catalysts, such as CrO_3 (5%) supported on finely divided silica-alumina (75%–90% silica) base in the Phillips process or molybdenum oxide (MoO_2) catalyst containing sodium or calcium metal or their hydrides as promoters in the Standard Oil of Indiana process were also developed. The Ziegler–Natta process uses the catalyst complex formed and dispersed *in situ* in *n*-heptane; the purified ethylene monomer, at a pressure of 20–40 atm, is allowed to polymerize at a relatively low temperature (70–100°C). The Phillips process and the Standard Oil of Indiana process for ethylene polymerization commonly use pressures in the range of 40–80 atm; temperatures between 130–150°C are used for the former, employing a hydrocarbon such as cyclohexane as the solvent, and the range of 220–250°C for the latter, also employing a hydrocarbon solvent.

When the reaction in the Ziegler process is completed to the desired extent, the catalyst is destroyed using water or alcohol with traces of an acid; the polymer that separates on cooling is collected by filtration or centrifugation, washed and dried and the unreacted monomer is recycled. The processes using metal oxide catalysts, are fixed-bed, moving-bed or slurry processes. On completion of the reaction to the desired degree of conversion, the polymer formed and remaining evenly distributed in the paraffin or cycloparaffin solvent in the high reaction temperature is separated from the catalyst fragments, recovered and isolated by cooling and/or solvent evaporation. Polymerization for the production of HDPE is understood to follow an anionic coordination mechanism.

Polyethylenes prepared or manufactured by these processes are by and large linear and are commonly called *high-density polyethylene* (HDPE); they are associated with the following characteristic property parameters: only 2–5 short branches per 1000 linear C–atoms in the chain molecules; melting temperature, 127–132°C; the degree of crystallinity of the bulk polymer; 80%–95%; tensile strength: 25–40 MPa and breaking elongation: 20%–200%. HDPE exhibits slightly better environmental stress cracking resistance (ESCR) than LDPE. The stress cracking problem is less critical with relatively high molecular weight (relatively low melt flow index, *MFI*) and high density materials.

The factors that contribute to the widespread applications of polyethylenes are low cost, easy processibility by a variety of techniques, excellent chemical inertness, non-toxicity, high electrical insulation, low-temperature toughness and flexibility, good clarity of thin films, heat sealability, water proofness and low water vapour permeability. Films commonly made from LDPE find application in heat-sealable packaging, waterproofing, water management including the lining of reservoirs and canals; pipes, insulated wires and cables, cans and carboys, mulches needed in irrigation and in fabric/paper coating and lamination. The limitations of polyethylene are its low melting/softening or heat-distortion temperature, proneness to environmental stress cracking resistance, low abrasion and scratch resistance, low tensile strength and high gas permeability.

(c) Chlorosulfonated Polyethylene (CSP)

The molecular symmetry of a high order for polyethylene (LDPE) may be broken by treating the polymer with chlorine in the presence of a small quantity of sulfur dioxide, thereby leading to the incorporation of some —SO_2Cl groups and Cl atoms along the polymer chain:

$$—CH_2—CH_2—CH_2—CH_2— + SO_2 + Cl_2$$

$$—\left\{\left[\begin{array}{c} —(CH_2)_n—\underset{\displaystyle Cl}{\underset{|}{CH}} \end{array}\right]_x —CH_2—\underset{\displaystyle SO_2Cl}{\underset{|}{CH}}—\right\}^y \qquad (3.1)$$

For an average, useful CSP, known by the trade name Hypalon, $n \geq 6$ and $x = 12$ provide a cross-linkable, rubbery product, where the $—SO_2Cl$ groups provide sites for cross linking or vulcanization in the presence of certain metal oxides, such as PbO, as the vulcanization agents. CSP rubber is superior to many hydrocarbon (diene) rubbers in ozone resistance, flex resistance, chemical resistance and heat aging.

(d) Chlorinated Polyethylene (CPE)

The chlorination of LDPE or HDPE is carried out at an elevated temperature in solution or in suspension to produce elastomeric products having chlorine content in the range of 40%–48% chlorine. Chlorination in an aqueous suspension at or above 120°C gives an elastomeric product that is conveniently vulcanized using peroxide vulcanizing agents.

(e) Linear Low-density Polyethylene (LLDPE)

The understanding of the dependence between different types and degrees of branching, the ranges of densities and other physical and mechanical properties of polymer systems like LDPE and HDPE has led to the development of a mechanistically linear polyethylene of a low or very low density range by copolymerization. The product, a linear copolymer of ethylene and a minor proportion of an α-olefin, such as 1-butene, 1-hexene or 1-octene, appears as a branched polyethylene having controlled and calculable proportions of uniform short (2-carbon, 4-carbon or 6-carbon) chain branching along the chain backbone. Uniform-length, short branches appear in the copolymer chain molecules at each point of the α-olefin comonomer incorporation. Such linear grades of polyethylene prepared using 4%–6% of the respective α-olefin comonomer is characterized by a density range (0.915–0.935 g/cm^3) and is generally called linear low density polyethylene (LLDPE) and those prepared using 10%–15% of the respective α–olefin comonomer exhibit a still lower density range (0.895–0.915 g/cm^3) and are referred to as Very Low Density Polyethylene (VLDPE).

Structurally, LDPE may also be viewed as a copolymer of ethylene and some proportion of assorted α–olefin due to significant numbers of long and short branches in its chain structure. HDPE, however, is by and large considered a linear homopolymer of ethylene.

LLDPE displays a superior balance[6] of toughness, stiffness, stress cracking resistance, clarity, heat sealing and electrical insulation properties compared to LDPE. Judicial blends of LDPE and LLDPE are being developed for more appropriate applications. The melting points of commercial grades of LLDPE are in between those of commercial LDPE and HDPE.

(f) Polypropylene and other Poly (α-olefins)

The polypropylene of commerce is *isotactic* and its synthesis became possible through the commercial utilization of coordination catalysts during 1957–60. The synthetic approach for polypropylene and other α-olefins is much the same as for the production of HDPE by the Ziegler process. The same reactor can be used to polymerize propylene or other α-olefins with some modifications, if necessary. In each case, particularly for propylene, the reaction condition should be so chosen as to overwhelmingly favour the formation of the isotactic isomer (>95%). The catalyst may be prepared from $TiCl_4$ and aluminium triethyl dispersed in *n*-heptane or solvent naphtha. Propylene is charged into the reactor under pressure. Nearly 80%–85% conversion is achieved in 8–10 h at 60°C. Hydrogen is used as a chain-terminating or chain length-controlling agent. Using propylene as the monomer, 3%–5% *atactic* polypropylene is commonly formed along with overwhelming formation of the desired isotactic polymer. After the desired conversion stage, the ingredients in the reactor are transferred into a flash drum to purge the unreacted monomer, which is recycled. The atactic polymer formed is soluble in the solvent naphtha, while the isotactic polypropylene separates out in the solvent medium on cooling. The solution part is decanted after centrifugation whereby most of the atactic polymer material is removed. The residue is then treated with methanol acidified with traces of HCl to decompose and dissolve the catalyst residues. The residual isotactic polypropylene is then centrifuged, washed, dried and collected.

Polypropylene (PP) of commerce is the lightest homopolymer known (density: 0.90 g/cm^3). With a melting temperature range of 170–175°C and glass transition temperature (T_g) of –10 to –20°C, it is viewed as an engineering plastic with good to excellent film forming capacity and a good degree of transparency in thin film form. PP turns somewhat brittle at temperatures ≤ 0°C. In this respect PP is inferior to PE (both LDPE and HDPE). PP, however, exhibits much improved hinge properties in box/luggage applications. The brittleness in PP is well mitigated by the block copolymerization of propylene with minor (5%–10%) proportions of ethylene. PP is generally free from environmental stress cracking problems. Its specific heat is lower than that of polyethylene but higher than that of polystyrene (Psty). All these hydrocarbon polymers (PE, PP, Psty, etc.) are very good to excellent as insulators. PP exhibits higher clarity (less opacity) than HDPE primarily because the amorphous and crystalline zone densities for PP are much closer than for the PEs. Due to its high melting point (170–175°C), PP is steam sterilizable. Biaxially Oriented Polypropylene (BOPP) shows a high degree of clarity; biaxial stretching also produces films/tubes of much improved tensile strength. Blended with other polyolefins, PP is viewed as a good polymer for textile coating and related applications. PP is more prone to oxidation than PE.

(g) Ethylene Copolymers

Among the copolymers of ethylene, those between ethylene and propylene and between ethylene and vinyl acetate have received widespread attention, development and application. Other important copolymers are those involving ethylene and a minor (3%–12%) molar proportion of an α-olefin (1-butene, 1-hexene, 1-octene, etc.), which are popularly known as LLDPE, with the use of upto 6% α-olefin or VLDPE, with the use of ≥ 10%–15% α-olefin. The LLDPE and related α-olefin copolymers of ethylene are produced by linear chain propagation mechanism using low pressure and coordination catalysts (Ziegler or Phillip's process) while copolymers of ethylene with vinyl acetate, maleic anhydride acrylic acid/methacrylic acid and some acrylate monomers are industrially produced by using free radical catalysts and high pressures.

(i) Ethylene-propylene (Copolymer) Rubbers (EPR) Stereo block copolymers of ethylene (10%–15%) and propylene (90%–85%) prepared by linear mechanism show much improved impact resistance and environmental stress cracking resistance. Though copolymeric in nature, they are essentially viewed as polypropylenes of minor chemical modifications. When the two monomers are used in nearly equal or 60 : 40 proportions, products better known as ethylene-propylene (EP) rubbers are formed on copolymerization. EP rubbers are more aging-resistant than diene rubbers including the synthetic varieties and natural rubber (NR), as EP is devoid of unsaturation; EP rubbers are conveniently vulcanized using peroxides (e.g. dicumyl peroxide) and heat. A low dose of sulphur (0.3 phr) is often used along with a peroxide (2–3 phr) as the curative and Ca-stearate is preferred to stearic acid as the activator. EPDM, a terpolymer of ethylene, propylene and a special diene (≤ 5%) such as ethylidene norbornene or dicyclo pentadiene is vulcanizable using accelerator-sulphur curative systems. It is used to in making coated fabrics, cables, tires and moulded articles.

(ii) Ethylene-vinyl Acetate (EVA) Copolymers Ethylene (E) and vinyl acetate (VA) are used as comonomers in different proportions to make a wide range of copolymers of special technical applications. Copolymers of low VA content (≤ 5%) are considered as modified low density PE. Their high surface tack, disadvantageous in conventional film applications, is taken advantage of in meat packaging as stretch films and cling wraps. With (a) 8%–12% VA, and (b) 15%–20% VA, or (c) between 25%–30% VA, the EVA copolymers are useful respectively as (a) wax additives and wax coatings, and (b) and (c) as hot melt adhesives suited in book binding industry. 28%–48% VA content makes different grades of soft EVA copolymers that are commonly useful as thermoplastic, mouldable elastomers of a wide range of properties; they are readily vulcanized using a suitable peroxide. Filled high-VA EVA has become useful in the footwear industry for making moulded soles and as a cementing agent, i.e. hot melt adhesive.

Higher VA content in the EVA copolymer produces (i) greater solubility, (ii) higher tack or cohesive power in the adhesive, (iii) better adhesion to polar surfaces and (iv) better low-temperature performance. EVA are available in wide range of melt (flow) indices (MFI) from 2–500; the higher MFI grades are desirable for processing as hot melt adhesives, due to their convenient low-range melt viscosity. For a balanced flow and viscoelastic character, it is common practice to blend EVA-copolymers of high and low MFI grades. The melting temperature and viscosity of such copolymers are controlled by molecular weight rather than by VA content.

Selected grades of hydrolyzed EVA, commonly designated as EVOH, offer excellent barrier properties with respect to gas (oxygen) permeability. Copolymers of ethylene with minor proportions of carboxylic monomers (maleic anhydride, acrylic acid or methacrylic acid) prepared by the high pressure process give products called 'ionomers' that are melt processed like thermoplastics but behave as thermosets (virtually like cross-linked materials) under ambient conditions. When treated with metal salts, oxides or alkoxides (Zn, Ca, Mg etc.), ionic metastable cross-links, stable under ambient conditions are readily established. The ionic cross-links become progressively weak with increased temperature and their influence practically disappears at the processing (melting) temperatures, thus permitting their reshaping under pressure. On cooling after reshaping, the ionic cross-links are regrouped and reformed and the thermo-hardened character reappears. The intermolecular salt bridging in ionomers appears as cross-links that are 'meltable'. Ethylene-ethyl acrylate (18% by weight) copolymer provides improved adhesion over PE surfaces and it shows better thermal stability than EVA copolymers.

3.4.3 Polystyrene and Styrene Copolymers

The styrene monomer is a colourless liquid with a characteristic fruity smell. It is prepared in the industry initially by ethylating benzene starting with ethylene and benzene using a Friedel–Crafts catalyst ($AlCl_3$) at 90°C and thereafter by the dehydrogenation of ethyl benzene at a much higher temperature using iron oxide or magnesium oxide as catalyst at nearly 600°C. Styrene isolated from the dehydrogenated product by distillation is stabilized by adding traces of a phenolic stabilizer such as t-butyl catechol and then stored in an air tight container in a cold storage for longer shelf life.

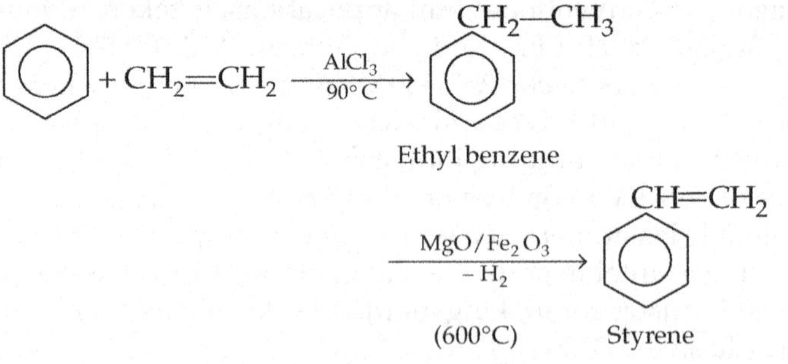

(3.2)

For thermoplastic applications, stabilizer-free styrene is polymerized in the absence or presence of selected comonomer(s) in bulk using a free radical catalyst such as a peroxide or an azo compound. However, polymers or, more appropriately, copolymers of styrene commonly meant for surface coating applications are polymerized (i) in solution using selected comonomers (acrylates, methacrylates, acrylonitriles etc.) all dissolved in an appropriate aromatic solvent following a solution polymerization technique, using organosoluble free radical catalysts or initiators, or (ii) in aqueous emulsion in combination with the selected comonomers using aqueous redox catalysts or initiators following an emulsion polymerization technique[1].

3.4.4 Synthetic Elastomers

(a) Styrene–Butadiene Rubber (SBR)

Styrene-butadiene rubber (SBR, 70–75 parts butadiene and 30–25 parts styrene) is prepared industrially by emulsion technique at a temperature of 40–50°C, or at much lower temperatures (5°C or even –10 to –15°C) using appropriate redox initiators (combination of a selected hydroperoxide oxidant and a metal ion or amine activator) and an appropriate dose of specified antifreeze ingredients in the initial emulsion polymerization system. Polymerization at a low temperature is a later development and products thus obtained are commonly known as cold SBR.

At the end of the polymerization process, when there is conversion of nearly 80%, a small dose of a shortstop or inhibitor (quinone) is added to arrest further polymer growth. Any unreacted monomer is then removed by blowing in steam. A requisite dose of an antioxidant is then dispersed in the latex. For producing oil-extended SBR, the right dose of an oil in water emulsion is added into the latex and if necessary, an appropriate grade and dose of carbon black may be dispersed into it at this stage to make black or oil-black master batch grades of SBR. The latex is finally coagulated, preferably in the form of small, porous granules, using salt (NaCl) and acid (H_2SO_4) solutions in succession; they are also used successively for washing the coagulants; alternatively, coagulation and washing are simply accomplished after the addition of a solution of animal glue into the latex. Alum solution may also be used as the coagulant.

The coagulated crumb is then separated from the serum and washed profusely with water; the coagulant's water-content is then brought down to ≤ 60% by draining on a vacuum or pressure filter. The amount of water is further reduced by roller pressure and then by drying in a large drier. The crumbs are then turned into bales by compression.

A solution process later developed for obtaining SBR is based on the use of butyl lithium as an anionic coordination catalyst at a low temperature (20–30°C). The solution process provides higher structural uniformity (1, 4-cis configuration for the butadiene units in the copolymer chain), much lower branching and higher economy through higher scope for ready oil extension and carbon black loading.

High-styrene, styrene-butadiene copolymers (30%–50% bound styrene) are useful as rubbers for designing tread compounds for tires with improved road grip and abrasion resistance when filled and reinforced with HAF grade carbon black. These high-styrene SBR grades are seldom used alone; they are blended with normal SBR or natural rubber (NR) for the production of relatively hard vulcanizates. Copolymers of styrene and butadiene (styrene content 80%–90%) are better known as high-styrene resins and are used in minor proportions as stiffening agents in NR- and SBR-based vulcanizates, along with some grades of acrylonitrile-butadiene-styrene (ABS) terpolymer or styrene-acrylonitrile (SAN) binary copolymers (having >75% styrene); they are useful as tough thermoplastics and some low molecular weight grades are useful as hot-melt adhesives.

SBR is notably superior to NR in aging and ozone resistance even though SBR develops cuts and cracks faster. SBR and NR have comparable hysteresis or heat build up, abrasion resistance properties and poor resilience characters. Other than its outlets in tire, hose and belting constructions, SBR is becoming popular in the production of floorings and coated fabrics.

(b) Polybutadiene Rubber (PBR)

High cis 1, 4 polybutadiene rubber (PBR), (>92% cis) is commercially prepared by the polymerization of butadiene in solution using selected coordination catalysts. Polymers of butadiene obtained by the emulsion technique using aqueous redox catalysts are very low in cis content (<10%). Oil-extended (20% oil) cis polybutadiene show vulcanizate properties that are substantially unchanged even without the requirement of carbon black filler adjustment. PBR shows high elasticity, rebound resilience and low hysteresis with somewhat improved resistance to oxidation. 20%–40% PBR blended with NR, SBR or NR-SBR blends offers excellent abrasion resistance to car and truck tires; a higher PBR content in the tire tread compound is undesirable as it decreases the skid resistance.

3.4.5 Other Synthetic Elastomers

(a) Polychloroprene Rubber (CR)

Polychloroprene is the world's first commercial synthetic rubber.

It is the polymer of 2-chloro 1, 3 butadiene ($CH_2{=}\overset{Cl}{C}{-}CH{-}CH{=}CH_2$); the relevant elastomer grades are commonly known as "neoprene". Emulsion polymerization is the prevalent technique used for polymerizing the chloro butadiene. The polymer has an overwhelmingly 1, 4 trans structure, which is primarily responsible for the crystallizable nature of the neoprenes (CR) and makes the chlorine atom reactive, being both tertiary and allylic and providing curing sites on the polymer.

The crystallizability and rate of crystallization of CR are reduced by the modification of the polymer structure by a minor proportion of a second monomer (possibly 2, 3 dichloro butadiene) during polymerization. The major classification and grading of neoprene is based on the extent of modification with sulfur or sulfur-donor compounds (disulfide accelerators) during manufacture; in the following classes of neoprene, neoprene GRT, neoprene WRT and neoprene WX, the extent of the second monomer incorporation is 10, 5 and 2.5%, respectively. The vulcanization of the neoprene is conveniently accomplished using 5 parts ZnO and 4 parts MgO per 100 parts of neoprene and heating to 140–150°C. The common sulfur-accelerator cure is ineffective and the use of small proportions (0.5–1.0 parts) of ethylene thiourea and antimony sulfide causes faster and effective press cure at 150°C. For optimum processibility, hardness and strength, fillers are commonly used, even though, in contrast with most synthetic rubbers, unfilled neoprene vulcanizates exhibit high tensile strength (3500–4000 psi) and relatively poor resilience. Using selected resins, plasticizers and oils is effective in controlling the crystallization tendencies of commercial neoprenes and the optimum vulcanization consequently effected is also helpful in this regard.

The major uses and applications of CR include heat-resistant and flame-resistant beltings, coated wires and cables, industrial hoses having good oil resistance, gaskets, O-rings and seals, rubber rollers, coated fabrics and adhesives. CR is somewhat inferior to nitrile rubber (NBR), i.e. butadiene-acrylonitrile (75 : 25) copolymer rubber in oil resistance, despite being substantially superior to other diene rubbers in this respect.

(b) Nitrile Rubber (NBR)

Nitrile rubber is a copolymerization product of butadiene and acrylonitrile (75 : 25), developed almost parallel to SBR in the 1930s. Copolymerization is almost exclusively done by the emulsion technique. Easy processing grades with predominantly linear structures, having little gel content were made available in the market by conducting the copolymerization at a low temperature (5°C). The relevant products are referred to as "cold" nitriles. Emulsion NBR prepared at a higher temperature range (25–50°C), called "hot" nitriles are relatively tough for having branch units and gels in the product. Acrylonitrile (AN) content in the NBRs commonly varies between 25%–40%, and even though the cost goes up with increased nitrile content, the relevant products are rendered more resistant to (non-polar) oil and solvent, gas permeation and somewhat inferior in low temperature flexibility.

Nitrile rubbers resist mastication, so, milling helps little to develop plasticity in them. The "cold" nitriles require little premastication and they develop less heat during mixing. Mills and internal mixers are commonly cooled to prevent scorching during the mixing of curatives or vulcanizing agents. Coated sulfur (e.g. $MgCO_3$ coated sulphur) may be used to minimize or overcome this mixing problem. High-boiling esters and polymeric plasticizers are commonly used to plasticize NBR to improve processiblity and confer a good balance in such properties as low-temperature flexibility, resilience, hardness, etc.

NBR is compatible with poly (vinyl chloride), PVC. A good balance of ozone and weathering resistance, high abrasion and oil resistances and improved flame resistance (for the use of chlorinated or phosphate plasticizer) may be derived from NBR-PVC blends on vulcanizing the elastomer component. NBR and NBR-PVC blends are useful in making gasoline hoses, lining for fuel storage tanks, and in wires, cables and sheaths for use in oil fields and oil installations. NBR finds special use in adhesive formulations and in the latex form, as impregnating agents for paper, textiles and leathers. Good abrasion resistance has widened the scope of NBR-PVC blends in footwear soling.

(c) Butyl Rubber (IIR)

Isobutylene can be commercially polymerized to high molecular weight polyisobutylene $[—CH_2—C(CH_3)_2—]_n$ using a cationic catalyst like boron trifluoride at −80°C. Though rubbery, polyisobutylene can not be vulcanized like NR or some other diene rubbers using accelerator-sulfur curative systems due to the absence of unsaturation in its repeat units. Using diene comonomer (2%–4%) such as isoprene along with isobutylene in methyl chloride diluent allows the formation of isobutylene-isoprene rubber (IIR), commonly known as *butyl rubber*. The polymerization reaction is fast and exothermic; so, cooling to maintain the desired low polymerization temperature (−65 to −80°C) is of utmost importance. Isoprene has a retarding effect on the copolymerization process and tends to limit the molecular weight of the copolymer. Using ≥ 5% isoprene is therefore unadvisable and unrewarding. The butyl rubber is obtained in a precipitated form of fine particles; the solvent is drained out and the polymer is admixed with an antioxidant in the wet condition, dried and baled.

Butyl rubbers show crystallizing tendencies on stretching to high elongations; they show poor compatibility with most diene and related rubbers and their low unsaturation comes in the way of satisfactory covulcanization. Their vulcanization is conventionally effected using quinone dioxime or other related dioximes along with 6 phr red lead. Butyl rubber offers more resistances to aging, ozone, abrasion, heat and chemicals than most other hydrocarbon rubbers. Its low gas permeability makes it suitable for excellent high-pressure, air-holding inner tubes in tires.

Butyl rubber, along with an appropriate dose of selected tackifiers, is suitable for pressure-sensitive adhesives. Butyl and more particularly the halo (chloro and bromo) butyl rubbers and relevant blends find use for making air bags, steam hoses, heat-resistant conveyor belts, inner linings of tubeless tires and for the white side walls of tires. The heat resistance of these rubbers is in the order butyl < chlorobutyl (1.5%–2% Cl_2) < bromobutyl (2.5%–3% Br_2).

(d) Polyacrylic or Acrylate Rubbers and Resins

Poly ethyl acrylate and other similar acrylates are soft and rubbery. Copolymers of ethyl acrylate (95%) and 2-chloroethyl acrylate or 2-chloroethyl vinyl ether (cure-site monomer) have grown to commercial production and are known as polyacrylic or acrylate rubber. This rubber is curable using linear aliphatic diamines and polyamines. A small amount of sulfur is used as an antiaging additive. Curing occurs by HCl elimination through a reaction between the Cl-containing cure-site monomer units and the amine curatives, and by the establishment of intermolecular amide linkages in part through the diamine/polyamine system via acrylate ester hydrolysis. The cured rubber is highly rated for sealants, gaskets, hoses and adhesives. In uncured form, this rubber is soluble in ketones, esters and alcohol-ester mixtures. The acrylate rubber vulcanizates are superior to nitrile rubber vulcanizates in heat and oil resistance, and in their resistance to ozone, sunlight and weathering.

A terpolymer (AEM) of methyl acrylate, ethylene and a cure site monomer acrylic acid, known in the trade as "Vamac" is a later development.[6]

Copolymers of acrylate (ethyl, 2-hydroxy ethyl, butyl etc.) and methacrylate (methyl and ethyl) monomers in the absence of small doses of acrylonitrile or styrene are obtained by emulsion polymerization or solution polymerization. Higher doses of acrylates make the copolymers soft and tacky while higher doses of methacrylates make them hard, strong, glossy and slightly brittle. The emulsion and solution polymers are useful as coatings, inks and adhesives. The suspension grade copolymers are suitable for moulding.

Poly (acylic acid) or more appropriately, zinc poly acrylate cements (bearing ZnO and minor amounts of MgO, SnO, SiO_2, Ca $(OH)_2$ and SnO_2 etc. as fillers along with poly (acrylic acid), swollen with water and giving 30%–50% of the polycarboxylate of molecular weight 30,000–50,000, act as good dental restoratives or dental cements that set by acid-base salt formation.

(e) Fluorocarbon Rubbers (FKM)

Copolymers of vinylidene fluoride ($CH_2{=}CF_2$) and chloro trifluoro ethylene ($CFCl{=}CF_2$) in 50:50 or 30:70 ratio were initially developed. The copolymerization of vinylidene fluoride and hexa fluoro propylene (the "Viton" elastomers from Du Pont) give perfluoro elastomers that are prepared by the emulsion polymerization technique. Batch or continuous processes have been developed. The latex is coagulated using hydrochloric acid and the coagulant is then washed and dried. The vitons are normally soluble in lower ketones and the doughs formed are readily spreadable over glass cloth to produce coated fabrics, which are useful as oil seals and gaskets with long service life at high temperatures ($\geq$100 h at 400°C and $>$ 5000 h at 200°C).

With excellent resistances to oxygen and ozone, oils, lubricants, hydrocarbon solvents, mineral acids, chemicals and to flaming and heat, fluorocarbon elastomers, are superior to most other commercial rubbers. Excellent low-temperature flexibility makes them continuously useful at as low a temperature as –30°C. The use of fluorocarbon elastomers is only justifiable if their high cost is not a constraint. Their major applications cover seals, gaskets and diaphragms, fire-resistant protective clothing from coated fabrics, wire insulation and cables.

Nitroso rubber, an alternating copolymer of trifluoro-nitrosomethane and tetrafluoroethylene of high molecular weight $[{-}N{-}\underset{\displaystyle CF_3}{\underset{|}{O}}{-}CF_2{-}CF_2{-}]_n$ has a highspecific gravity (1.9). It is also curable using diamines and polyamines. Nitroso rubber is more resistant than the fluoro carbon elastomers to oils, solvents, acids and alkalis, chemicals, strong oxidizing agents and to oxygen and ozone attack.

(f) Poly Tetrafluoroethylene (PTFE)

Commercial polymerization of tetrafluoroethylene ($CF_2{=}CF_2$) is essentially aqueous polymerization accomplished by using free radical initiators ($S_2O_8^=$, H_2O_2), maintaining an elevated pressure. Polymers are obtained in powder/granular forms or in aqueous dispersion. PTFE shows moderate tensile strength (2500–4000 psi), excellent heat resistance and high meting point (327°C) with very high melt viscosity. Dispersions are used in making coated items (non-stick PTFE coated pans and kitchenware). Powder metallurgy or ceramic processing are employed to obtain moulded objects, seals, films, lab-equipments, electrical gadgets and machine components, etc.

(g) Polysulfide Rubber

Polysulfide rubbers, popularly known in the trade as "Thiokols," are step growth polymers obtained from the reaction between dichloro alkanes and sodium polysulfide.

$$n\,RCl_2 + n\,Na_2S_x \rightarrow (\text{—}R\text{—}S_x\text{—})_n + 2n\,NaCl \qquad (3.3)$$

The high content of linear or cyclic sulfur linkages in the polymer chains render the products desirably rubbery.

The polysulfide rubbers have good resistance to oxygen and ozone attack and they offer excellent resistance to gas permeability, oils and solvents. However, they have a disagreeable odour and offer poor processing characteristics and mechanical properties. Major applications include tank lining, hoses for gasoline, and gaskets, sealants and diaphragms.

A combination of zinc oxide, stearic acid and the usual rubber accelerators (MBTS, TMTD, etc.) or p-quinone dioxime without sulphur acts as an effective curative for polysulfide rubbers. Organic peroxides or metal peroxides (PbO_2) are also useful curatives. Polysulfide rubber vulcanizates show notable improvement in strength with carbon black loading to the extent of 40–60 phr. The liquid thiokols burn readily, producing large volumes of acidic gas when mixed with strong oxidizing agents.

(h) Cyclized Rubber and Chlorinated Rubber

Cyclization of natural rubber (NR) is readily achieved by treating or mildly heating the rubber in the presence of mineral acid or Lewis acid catalyst, such as H_2SO_4, $SnCl_4$, $TiCl_4$, BF_3, etc. The cyclized rubber bears much lower degrees of unsaturation and the product shows resinous characteristics. A theoretical monocyclic rubber is expected to exhibit a drop in unsaturation to 57% of the initial NR. But for many commercial products, the unsaturation drops to 20%–30% of the initial rubber, thereby indicating the formation of some bicyclic or even polycyclic structures. A typical bicyclic structure[5] combining three successive 1, 4 polyisoprene chain units is shown is Eq. 3.4:

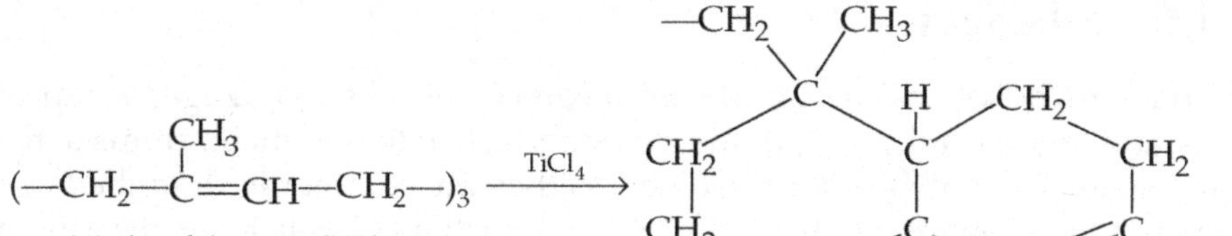

1.4 *cis-polyisoprene (NR)*

(3.4)

Cyclized rubber (bicyclic structure)

The current practice is to make cyclized rubber directly from the NR-latex. In the presence of selected stabilizers, latex may be made acidic without coagulation. A cyclized rubber masterbatch is prepared by heating such specially stabilized latex with concentrated H_2SO_4, and mixing the relevant product, still in dispersion form with untreated latex containing a comparable quantity of rubber and then coagulating the mix. The coagulum is washed and dried and the masterbatch is ready for commercial use, particularly as an ingredient for making stiff vulcanizates.

NR can be readily halogenated. Only chlorination has been developed to a commercial scale. Direct chlorination with chlorine is associated with both addition and substitution reaction and HCl is evolved as a byproduct. For additive chlorination, a maximum of 5.1% chlorine is attainable, but the commercial products bear 60%–68% chlorine and they are resinous in character. It is commonly believed that a good degree of cyclization also occurs on chlorination. Rubber, cut into small pieces, is dissolved in CCl_4 in the presence of a small amount of benzoyl peroxide, which acts as a depolymerizing agent and lowers the solution viscosity. A fairly uniform chlorination is achieved by spraying the solution at the top of a chlorinating tower in which a stream of chlorine, let in at the bottom, is allowed to ascend. Droplets of the chlorinated product collect at the bottom, which is finally degassed to remove any excess chlorine. The solution is sprayed into a steam chamber where the heat of the steam causes rapid volatilization in CCl_4; the solvent, CCl_4, is then recovered for reuse. The wet, chlorinated mass collected at the bottom is washed, dried, milled if necessary and stored. Chlorinated rubber is resistant to many chemicals, including concentrated nitric acid. It is, however, soluble in a wide range of solvents and is variously useful in formulations of paints, lacquers, adhesives, and printing inks. High chlorine content confers prominent flame retardancy on the product.

NR hydrochlorination in benzene solution using HCl gas is associated with some cyclization and produces a resinous product. Films made from hydrochlorinated NR solutions enriched with the right doses of plasticizers or stabilizers are used for making laminates of paper or other plastics such as cellulose acetate in film form. Such laminates are good as heat-sealable packages for dry food, cosmetics, shampoo, etc. The NR-hydrochloride is also useful in adhesive formulations and acts as a bonding agent.

3.4.6 Silicones

Silicones are polymers having silicon-oxygen (—Si—O—) linkages in repeat units, and organic (alkyl/aryl) side groups attached to the silicon atoms in the main chain. The poly organosiloxanes, as they are commonly described, are prepared by allowing chloro organosilanes such as $Cl_2SiR_1R_2$ to react with water to form the corresponding hydroxy compounds on hydrolysis; these compounds subsequently condense with each other, leading to the formation of polymer molecules.

$$n\,Cl-\underset{R_2}{\overset{R_1}{\underset{|}{\overset{|}{Si}}}}-Cl \xrightarrow[-HCl]{H_2O} n\,HO-\underset{R_2}{\overset{R_1}{\underset{|}{\overset{|}{Si}}}}-OH \xrightarrow{-H_2O} \left(-\underset{R_2}{\overset{R_1}{\underset{|}{\overset{|}{Si}}}}-O-\right)_n \tag{3.5}$$

Only dichloro-alkyl or dichloro-aryl silanes yield linear polymers; monochloro silanes fail to produce any polymer and trichloro-monoalkyl (or aryl) silanes yield network polymers. The silane intermediates R_n Si X_{4-n} $(n = 0-4)$, where X is a hydrolyzable group or atom such as alkoxy/aryloxy or chlorine, may be made by direct synthesis involving reactions between alkyl/aryl halides and elementary silicon :

$$2CH_3\,Cl + Si \xrightarrow[Cu]{280\,^\circ C} Si\,(CH_3)_2Cl_2 \tag{3.6}$$

Dimethyl dichloro silane is the major (70%–75%) product under the most favourable conditions; the minor constituents in the crude product are methyl trichloro silane (8%–10%) and trimethyl chloro silane (5%–6%). They are separated by fractional distillation.

Silicone fluids (oligomers) are usually an equal mixture of low polymers and cyclic condensation products of silanols obtained by the hydrolysis of a predetermined mixture of chlorosilanes in excess water by vigorous agitation. To stabilize the viscosity and to obtain higher proportions of linear polymers, the fluid is required to be equilibrated by heating with dilute H_2SO_4. Monofunctional trimethyl chloro silane is used as a chain length-controlling or 'blocking' agent. The equilibration may take several hours at 100–150°C. The oil is separated from the aqueous layer and neutralized by washing with sodium carbonate solution. Silicone fluids and polymers of different viscosity grades are all colourless, odourless, water repellent and non-toxic products of high thermal stability, showing very narrow deviations in physical properties, e.g. viscosity, over a wide range of temperature (–70 to 200°C). They are used as polish additives; as mould release agents in plastics and rubber moulding operations; in die casting of metals and as greases. They are also useful as antifoam additives, in drug formulations and surgery, as additives in paints and other surface coatings and as water-repellent surface finishes in textile, paper and leather industries.

Silicone elastomers or rubbers are very high molecular weight grades of linear siloxane polymers prepared from the difunctional monomer $(CH_3)_2\,SiCl_2$ or from its cyclic tetramer, often mixed with minor proportions of dichloromethyl phenyl silane. The silicone elastomers are commonly vulcanized using peroxides such as benzoyl peroxide. Filler incorporation is particularly necessary for obtaining relatively strong, stiff vulcanizates; unfilled vulcanizates are soft and weak. Silica fillers are preferably used for reinforcement. Silicone elastomers with vinyl side groups (0.02–0.5 mole percent of other (methyl/phenyl) organic side groups) may be more readily cured using less reactive peroxides such as dicumyl peroxide to produce more elastic vulcanizates. The silicone rubbers are much more expensive than other synthetic rubbers and they offer inferior physical and mechanical property combinations than those exhibited by other synthetic rubbers. Silicone rubbers offer good resistance to weathering, oxygen and ozone attack and to the action of lubricating oils. They show high compression set (20%–50% in 24 hours at 150°C), very high air permeability and excellent low temperature (–80°C) flexibility. As special purpose rubbers, they are useful in wire coating and cable insulation that demands fire resistance and also as sealants and gaskets. The cold-cure silicone elastomers are suitable for sealing, caulking and encapsulation. Their physiological inertness make silicone rubbers suited for use as sterilizable blood transfusion tubing and closures of antibiotic containers.

Silicone resins are made from a suitable blend of chlorosilanes, giving an average R/Si ratio ranging between 1.2 : 1 to 1.6 : 1 by acid hydrolysis, avoiding overheating. The polymer formed in liquid (oily) form is separated from the aqueous layer and neutralized with a mild alkali wash. The molecular weight is then allowed to advance without cross-linking by heating or bodying the fluid system in the presence of a metallic soap to a desirable viscosity. Cross-linking the resin by a peroxide is achieved *in situ* in the finished product on heating. The silicone resins find applications in electrical insulation in the form of glass cloth laminates and insulating varnishes. Brushing, spraying and dipping followed by heating for solvent evaporation and cross-linking (where necessary) are common application techniques employed. The resins are useful in special purpose water-repellent industrial paints and as encapsulating agents.

3.4.7 Polyurethanes (PU)

Polymers prepared by the interaction of isocyanate groups of di- or polyisocyanates and hydroxyl groups of monomeric or polymeric diols or polyols are commonly known as polyurethanes (PU). The reaction between hexamethylene diisocyanate and 1, 4 butane diol yields plastic-grade and fibre-forming polyurethane (Perlon U). The reaction is essentially a rearrangement (rather than condensation) polymerization characterized by a step-growth mechanism:

$$n\,\mathrm{OCN.\,(CH_2)_6.\,NCO} + n\,\mathrm{HO.\,(CH_2)_4.\,OH} \rightarrow [\mathrm{—OOCNH \cdot (CH_2)_6 \cdot NHCOO\,(CH_2)_4—}]_n \qquad (3.7)$$

Different ranges or grades of polyurethanes are available for use as resins, rubbers, fibres, adhesives and surface coatings. Their applications as flexible or rigid foams are, however, of special importance.

The polyurethane (Perlon U) tends to decompose at temperatures ≥ 220°C; so overheating should be avoided. The PU has lower moisture absorption and a lower melting point than the corresponding polyamide (nylon). Lower softening temperature limits the use of PU (Perlon U) as fibres and textiles. The filaments/fibres find some use as bristles, monofilaments, multipurpose nets, sieves and filter cloth. The polymer is otherwise useful as a thermoplastic resin and is readily processed by injection moulding and extrusion techniques. The moulded and formed objects exhibit good dimensional stability and good retention of electrical properties even under humid conditions. Much like the nylons, polyurethanes, too, readily dissolve in formic acid and phenols.

Polyurethane rubbers of different types and grades are usually the products of chain extension reactions involving linear aliphatic polyesters or polyethers with terminal —OH groups and diisocyanates (toluene diisocyanate, 1, 5 naphthalene diisocyanate, etc.). The urethane linkages established during chain extensions may subsequently react with excess diisocyanates at a higher temperature during moulding to establish cross linkages. The PU rubbers show good tensile strength (~ 5000 psi), higher than any other rubber, and excellent tear and abrasion resistances. They also exhibit excellent resistances to oxygen and ozone attack. They are, however, less resistant to acids and alkalies and to steaming (hydrolytic degradation) than most other synthetic rubbers. They are particularly useful as oil seals, footwear soles and heels, in the construction of conveyer belts, printing rollers, diaphragms and industrial hoses, and in fabric coating, waterproofing items, tank-lining in chemical plants, pump impellers and in oil-resistant cable construction (sheathing).

3.4.8 Polyester (Alkyd) Surface Coating Resins as Base Materials for Oleoresinous Varnishes

One major application area of synthetic polyesters relates to surface coatings. They are also commonly referred to as *alkyd* resins or *glyptal* resins.[6,7] They are primarily synthesized by reacting combinations of diols or polyols (alcohols) and dicarboxylic acids in the presence of fatty acids, oils or some modifying resins at high temperatures (200–270°C) with or without a mineral acid as catalyst. The alcohols more commonly used are glycerol, ethylene glycol, 1, 3 propylene glycol, pentaerythritol, dipentaerythritol and sorbitol. The common acids in use are phthalic and maleic anhydrides and sebacic acid, and a host of saturated/unsaturated monocarboxylic acids and related carboxyl bearing resins used as modifying or blocking agents. Removing the byproduct water by vacuum helps the polycondensation (step growth) reaction move forward.

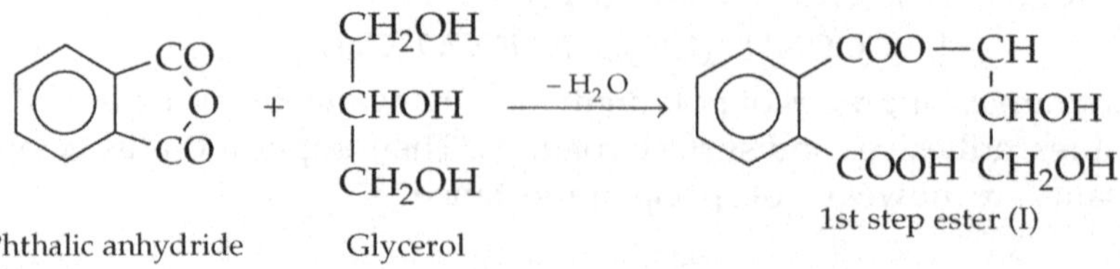

(3.8)

Phthalic anhydride Glycerol 1st step ester (I)

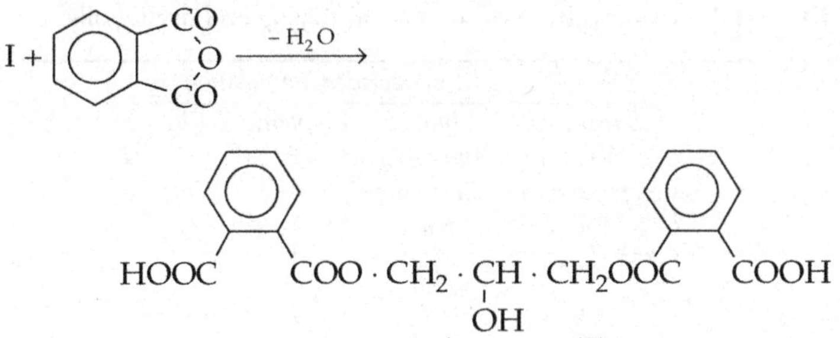

(3.9)

2nd step ester (II)

$$\text{II} + \begin{array}{c}CH_2OH\\|\\CHOH\\|\\CH_2OH\end{array} \xrightarrow{-H_2O} \begin{array}{c}\text{Long chain}\\\text{polyester (alkyd)}\\\text{with branching}\\\text{(III)}\end{array} \xrightarrow{-H_2O} \begin{array}{c}\text{Network}\\\text{polyester}\\\text{(alkyd)}\\\text{(IV)}\end{array} \qquad \textbf{(3.10)}$$

The long chain reaction product upto the stage of formation of (III) remains soluble and fusible; the product (IV) obtained in the final stages is a network polymer that is insoluble and infusible. The products (II and III) in the intermediate stages of growth are polyfunctional with the potentiality of further chain growth by a step-reaction (condensation) mechanism.

The scope for checking gelation through premature cross linking and desired enhancement of internal plasticization is simultaneously accomplished by initially blocking one alcohol group of the trihydric alcohol glycerol (R $(OH)_3$) with a monocarboxylic acid such as a long-chain (18-carbon) fatty acid (R′COOH), thus rendering the first step product (I) by and large bifunctional:

$$\underset{\text{(Glycerol)}}{R(OH)_3} + \underset{\text{(Monocarboxylic acid)}}{R'COOH} \xrightarrow{-H_2O} \underset{\text{(Bifunctional intermediate)}}{\begin{array}{c}HOROH\\|\\R'COO\end{array}} \qquad (3.11)$$

The internal plasticization effect is prominent due to the use of selected 18-carbon fatty acids as the blocking agents. Using a non-drying (saturated) fatty acid would produce non-drying resins. Using drying fatty acids, however, ultimately results in the formation of interchain oxidative cross linkages on baking, stoving or air drying, and such resins are commonly referred to as *drying-type* resins.

(a) Drying, Semi-drying and Non-drying Oils

The polyester resins for surface coating are mostly glyptal (glycerol-phthalic anhydride-based) resins[7] usually modified with a semi-drying or drying oil as listed in Table 3.2.

TABLE 3.2 Composition of some semi-drying and drying oils

Oil	*Glyceride Composition %*				
	Saturated and mono-olefenic acids	*Linoleic and other diolefenic acids*	*Linolenic acid*	*Eleostearic acid*	*Licanicacid*
Corn	58	42	—	—	—
Soybean	40	56	4	—	—
Linseed	31	24	45	—	—
Dehydrated castor	16	84	—	—	—
Perilla	15	38	47	—	—
Tung	20	—	—	80	—
Oiticica	18	—	—	—	82

(b) Fats and Oils

Animal and vegetable fats and oils primarily consist of mixtures of mixed triglyceryl esters of long-chain fatty acids as given by structure (V) below:

$$\begin{array}{l} CH_2-OOCR_1 \\ | \\ CH-OOCR_2 \\ | \\ CH_2-OOCR_3 \end{array}$$

(V)

Here, R_1, R_2 and R_3 are usually 17-carbon linear saturated or unsaturated hydrocarbon groups. Oils may be *non-drying* (containing mostly saturated acid esters), *semi-drying* (containing significantly mono-olefenic acid esters too) or *drying* (containing primarily di-and triolefenic acid esters).

The unsaturated acids that commonly form vegetable/animal fats and oils are shown in Table 3.3.

TABLE 3.3 Major unsaturated acids that largely constitute fats and oils

Acids	*Chemical Structure*
Oleic acid	$CH_3\,(CH_2)_7\,CH{=}CH\,(CH_2)_7\,COOH$
Ricinoleic acid	$CH_3(CH_2)_4CH_2CH(OH)\,CH_2\,CH{=}CH(CH_2)_7COOH$
Linoleic acid	$CH_3(CH_2)_4CH{=}CH\,CH_2\,CH{=}CH\,(CH_2)_7\,COOH$
Linolenic acid	$CH_3CH_2CH{=}CH\,CH_2\,CH{=}CHCH_2\,CH{=}CH(CH_2)_7\,COOH$
Eleostearic acid	$CH_3(CH_2)_3CH{=}CH\,CH{=}CH\,CH{=}CH(CH_2)_7\,COOH$
Licanic acid	$CH_3(CH_2)_3CH{=}CH\,CH{=}CH\,CH{=}CH(CH_2)_4CO(CH_2)_2COOH$

Drying an oil in air is accompanied by the following changes:

(a) oxygen absorption,
(b) lowering of iodine number,
(c) lowering in unsaturation (diene number),
(d) decrease in solubility,

(e) increase in viscosity and
(f) increase in average molecular weight

All contributing to the change of the polyunsaturated oil into a tough, somewhat rubbery solid. Drying is actually an oxidative polymerization process leading to gelation. Saturated oils do not dry. The appearance of unsaturation in conjugation results in a more pronounced and fast drying character. Upon hydrolysis, drying oils yield nearly 90% or more of unsaturated acids which are by and large 18-carbon atom linear acids with a striking uniformity of having the presence of the grouping —$(CH_2)_7COOH$ in most of them.

(c) Bodied oil and Drying Mechanism

Most oleoresinous varnish mixes contain several resinous and polymeric constituents and drying oil, which are cooked together and heated to 250–300° C. Polymerization and adduct formations with the resins take place even in the absence of oxygen. Thermal polymerization (boiling), or "heat bodying" as it is called, is an integral part of the oleoresinous varnish technology. A modified Diels-Alder reaction is involved in the process, where G represents the glycerol end of the fatty acid chain.

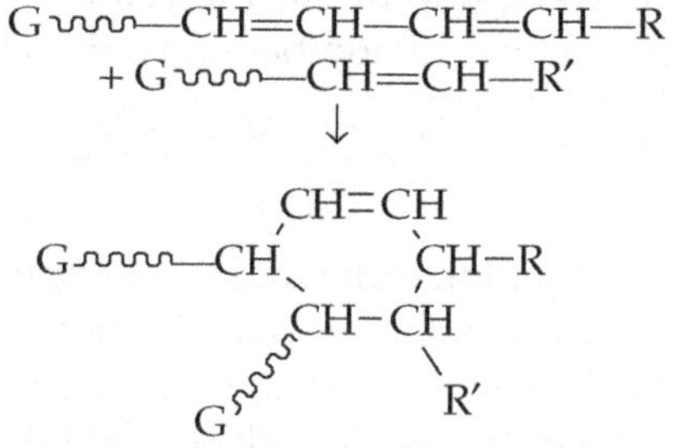

(3.12)

For a non-conjugated system of double bonds as originally present in linoleic and linolenic acid units, a rearrangement to a conjugated structure precedes of cyclization and cross-linking, as shown in Eq. 3.12. Oxidative polymerization (drying) of bodied oil spread out in a film form proceeds faster in the presence of selected oil-soluble metallic soaps as catalysts called driers (cobalt or copper naphthenate, lead rosinate, etc.) which help in transforming the applied coating to a solid, dry film. The dried film is a mutaplast of prominent hydrocarbon character and very poor symmetry, thus accounting for its flexibility, toughness and somewhat rubbery character.

Oils having a high degree of conjugated unsaturation, and hence a fast drying character, yield the densest molecular structure; they tend to discolour to a considerably greater extent than what results from using less conjugated oils. So, in clear varnishes, white paints and enamels, some drying speed must be sacrificed for improved colour and shade of coatings by incorporating soybean oil, dehydrated castor oil (DCO) or other relatively slow-drying oils into the mix.

Oil bodying is usually associated with a certain amount of glyceride splitting, releasing some free fatty acids, which then remain an integral component of the bodied oil; they get bound by reacting with the basic pigments and resins used. The presence of free acid is expressed by *acid number*. Free acid apparently retards the thermal polymerization process; if kept in control, its presence may in fact turn advantageous in respect of controlling, delaying or minimizing the process of sudden gelation during varnish cooking.

To arrive at a proper balance of properties, it is necessary to add some resin, natural (rosin, limed rosin or ester gum, shellac, copal or dammar) or synthetic (phenolic, urea melamine resin, alkyds). The resin content needs to be optimized to avoid the development of brittleness. The *oil length* of an oleoresinous varnish is the number of gallons of drying oil per pound of resin. A long oil varnish is tough and durable and indeed more suitable in outdoor applications. A short oil varnish is harder and more brittle and takes a higher polish, which makes it suited as a furniture finish.

Natural resins contain mono- and di-carboxylic acids as well as hydroxy acids and esters. They require to be given a vigorous cook to make them compatible with and soluble in varnish oils. Cooking thus enables intermolecular esterfication between the hydroxyl group of one molecule and the carboxyl group of another, leading to a drop in acid number and a rise in viscosity and average molecular weight.

(d) Ester Interchange

While cooking varnish mixes containing different oils or triglyceryl esters of different compositions in the presence of a resin at about 220–240°C, extensive ester interchange reactions[6,7] between the triglycerides, and between the resin (acid and ester) used and thermally-split triglycerides take place; the net result is a uniformity of molecular composition that was not present in the raw, uncooked mix. After this uniformity is achieved by ester interchange or interesterification reaction, the kettle temperature is raised to 250–300°C to extend the thermal polymerization of the glycerides to the desired extent avoiding gelation and undue glyceride splitting.

(e) Different Types of Alkyds

Alkyds that are commonly known as *plasticizing resins* are primarily formed from glycerol, phthalic anhydride or sebacic acid and a non-drying oil such as castor oil or dehydrated castor oil (DCO). They are chiefly used in lacquers to impart flexibility; they are applied commercially as 50–80% solution in butyl acetate, toluene or xylene. They lack hardness and are relatively slow drying; they are compatible with a host of hard resins for which they act as good plasticizers. They impart flexibility, improved stability and improved adhesion to metals when mixed in a cellulose nitrate lacquer formulation. They are property improvers in hard-baking enamels of white and pastel colours, particularly when the alkyd is combined with alcohol-soluble urea-aldehyde (u-f) or melamine-aldehyde (m-f) resins.

Alkyds, better known as drying resins, are made from phthalic anhydride, glycerol and drying oils or drying oil acids. Those high in fatty acid content are soluble in petroleum (aliphatic) thinners, but those having higher phthalate content need aromatic solvents and have low tolerance for aliphatic thinners. They are commonly based on the use of linseed oil or tung oil as the drying oil. They possess some dark shades and offer an excellent combination of properties but lack hardness when used by themselves. However, in combination with phenol-aldehyde resin or more appropriately, urea/melamine resins or maleic-rosin adduct resins, they exhibit much improved hardness and adhesion. Such drying oil resins based on pentaerythritol or dipentaerythritol as the alcohol component along with phthalic anhydride in the presence or even absence of maleic anhydride offer all the desired advantages of colour, flexibility, adhesion, hardness, gloss and quick drying.

Alkyds that are commonly referred to as hard resins are primarily based on the use of maleic anhydride, glycerol, excess rosin and drying oil. All grades of this resin have high melting points (120–150°C), are light in colour, have excellent hardness and good adhesion. They are soluble in aromatic hydrocarbons, turpentine, hydrogenated petroleum naphtha and ethyl acetate. They are widely used in both varnish and lacquer coatings where good adhesion, hardness, pale colour and gloss are required.

3.4.9 Phenolic Resins

Phenols in general readily react with aldehydes, forming resinous products. Common (the simplest) phenol and formaldehyde are most widely used for making industrial phenolic (phenol formaldehyde) resins. During phenol-aldehyde reactions, water is evolved as a byproduct but the phenolic-OH group remains unaffected. With the two ortho positions and one para position open or remaining unblocked, one may obtain linear or cross-linked (network) resinous polymers, depending on reaction conditions and the time of reaction; however, with one of the two ortho positions or the only para position blocked, phenol-aldehyde condensation can produce only linear resins or polymers under all conditions. Ortho- and para-hydrogens are the reactive centres in phenol in such condensation reactions; phenol is considered trifunctional, while formaldehyde ($CH_2{=}O$) in aqueous solution (formalin) is viewed as its hydrated product methylene glycol ($HO{-}CH_2{-}OH$) which is characteristically bifunctional.

Chemistry of Resin Formation

The phenol–formaldehyde resinification system is a tri-bifunctional one; the resinification proceeds by a step-growth polycondensation mechanism, invariably leading to the ultimate formation of (space) network polymers whose network density or cross-link density would be contingent upon the initial phenol aldehyde molar ratio or on the disturbance of the said ratio at a later stage of resin formation[6,7]. The first step in the resinification process relates to the formation of *monomethylol phenol* as in Eq. 3.13:

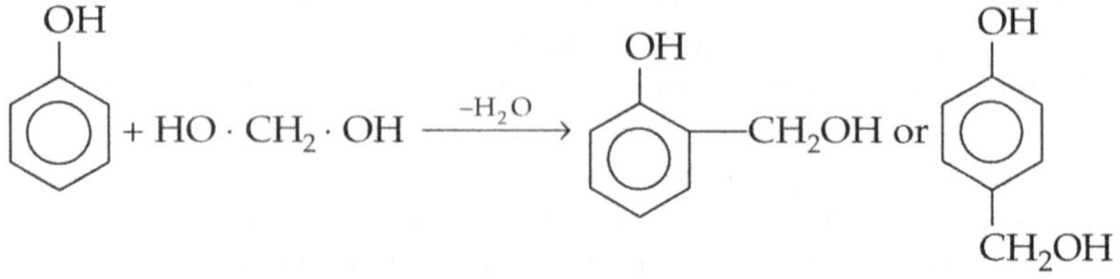

(Phenol) (Methylene glycol, i.e. formalin) (Monomethyl ol phenols) (3.13)

The reaction is catalyzed by both acids and bases. The basic catalyst (Na_2CO_3, NaOH, NH_4OH, etc.) and higher molar proportions of formaldehyde favour the formation of *di- and tri-methylol phenols* in the subsequent steps, as in equation (3.14).

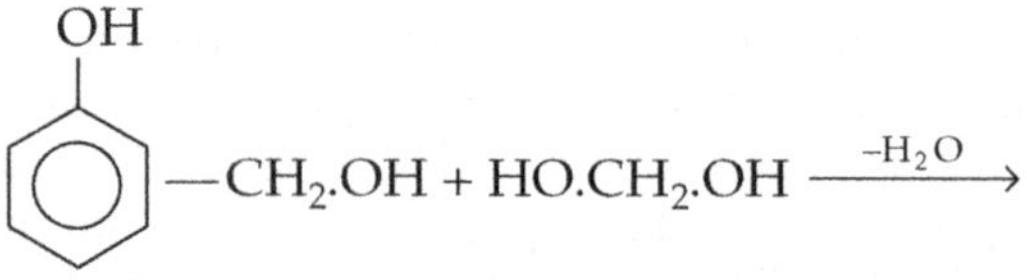

(Monomethylol phenol) (Methylene glycol, i.e. formalin)

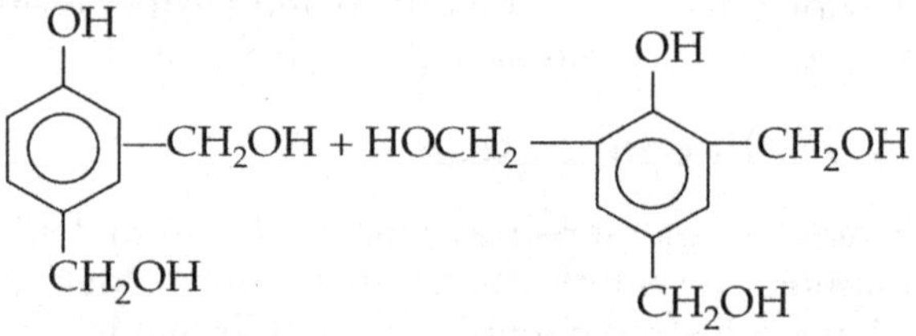

(Dimethylol phenol) (Trimethylol phenol) (3.14)

The initial reaction products, or the resinoids and the reactants, further react among each other, leading to chain extension and branching with the consequent formation of resinoids of progressively increasing molecular complexity and molecular weight, and lower solubility or insolubility in alcohol.

However, acids are more powerful catalysts. Acid catalysts (mineral or organic acids) and higher molar proportions of phenol favour an immediate reaction involving two phenol molecules for each formaldehyde molecule, resulting in the production of 2, *4-dihydroxy diphenyl methane* as the primary isolable product in the early stages. The early stage resin intermediates readily react with more formaldehyde and phenol in succession to form larger monomethylol derivatives of progressively higher molecular weight, till the resin formed practically becomes methylol-free and separates at the bottom of the reactor, leaving a clear, aqueous layer on top. The resin mass isolated by decantation or removal of the top aqueous layer can be further heated to drive out the absorbed water and get a resin melt that solidifies to a brittle resin mass on cooling; the brittle resin so obtained has a molecular weight of

600–700, and is commonly known as *novolac*. Novolacs are fusible and permanently soluble in alcohol.

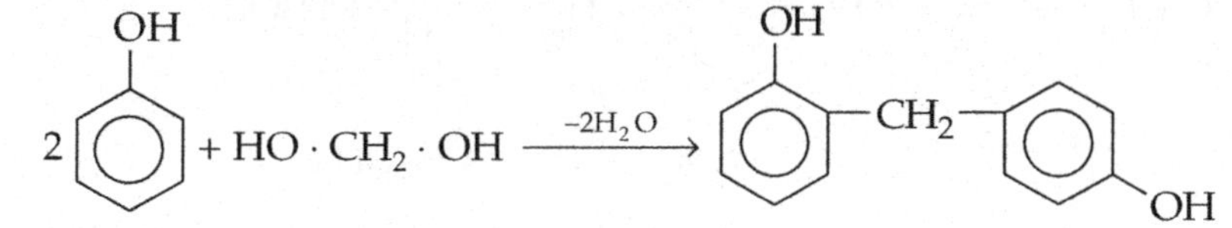

(3.15)

(Phenol) (Methylene glycol, i.e. formalin) (2, 4-dihydroxy diphenyl methane)

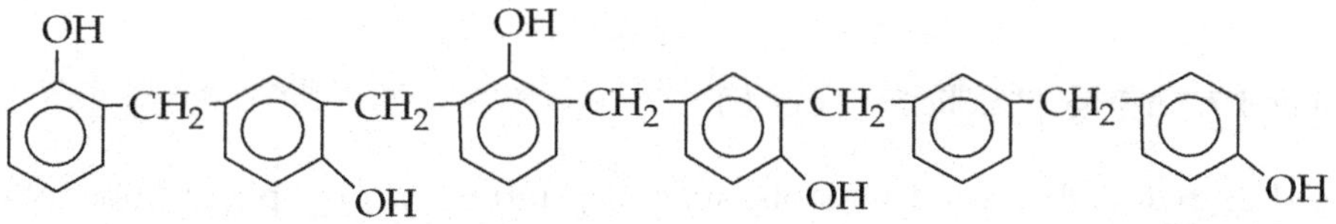

Average novolac molecule

A typical novolac molecule contains about 6 aromatic rings linked through methylene linkages and have practically no methylol group in the resin molecule. If additional formaldehyde is made available and the resin is thermally treated, cross linking through methylolation and intermolecular methylene bridge (—CH_2—) formation with liberation of water as byproduct occurs readily, ultimately forming an insoluble and infusible brown to dark network of cross-linked resinous products.

Cross-linking the novolacs in the second step (novolac technology being a two-step process) in the presence of additional formaldehyde is better accomplished using a basic catalyst. In fact, using 10%–15% of hexamethylene tetramine, simply known as hexa, fulfils the dual purpose under heated conditions (>100°C) as it thermally decomposes to liberate ammonia (which serves as the base catalyst) and formaldehyde (cross linker) by reaction with the moisture absorbed in the resin system and deriving it from the environment.

$$\underset{\text{(Hexa)}}{(CH_2)_6 N_4} \xrightarrow{+6H_2O} \underset{\text{(Ammonia)}}{4\,NH_3} + \underset{\text{(Formaldehyde)}}{6\,CH_2O} \qquad (3.16)$$

Base catalyzed phenol-formaldehe (p-f) resinification leading to the formation of a cross-linked product is described and recognized as a one-step process even though three successive stages of growth to more complex and bigger molecules exist. The first stage resin is soluble in water and alcohol; it is called the A-stage resin or resole and being rich in methylol groups and having a molecular weight of 300–500, may be described by a branched structure as in Fig. 3.7.

On further heating, resoles (soluble in alcohol) grow in size in all directions, forming a highly branched and mildly cross-linked intermediate (B-stage) product called *resitol* (insoluble in alcohol, but partly or fully soluble in acetone). On further heating, resitols grow into a giant network polymer called the C-stage resin or *resite,* which is completely insoluble and infusible. Resins

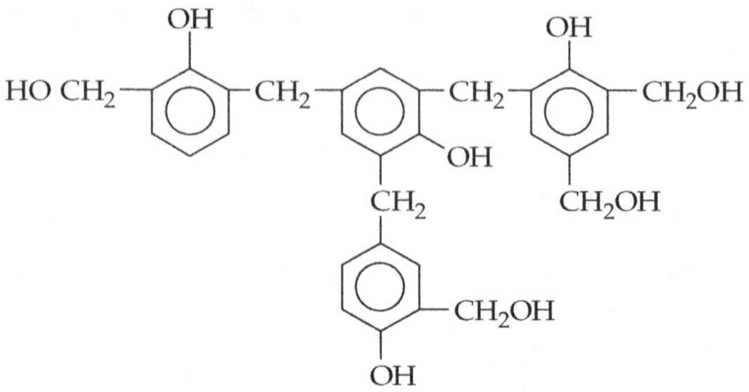

FIGURE 3.7 A-stage plenolic resin (resole)

based on para substituted phenols, such as p-tert butyl-or p-phenyl phenol react with formaldehyde to form *oil-soluble* grade phenolics that are used in surface coating formulations.

Phenolic resins are widely used:

(a) as adhesives to make paper, fabric and plywood laminates for mechanical, decorative and electrical uses
(b) as binding resin for making moulded objects, brake linings, gears, bearings and bushing, grinding wheels and
(c) for making modern heat- and corrosion-resistant refractory bricks, mortars, slide plates and linings

3.4.10 Amino Resins

Resins or plastics derived from reactions between amines or amides and aldehydes are commonly referred to as *amino resins* or *amino-plastics*. Of the various amino resins prepared and used, the *urea-formaldehyde* (*u-f*) resins are the most important in many respects; next in importance are the *melamine-formaldehyde* (*m-f*) resins.

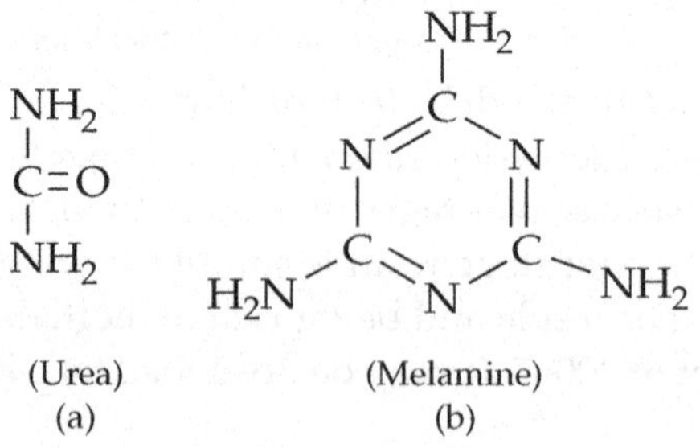

FIGURE 3.8 Structure of (a) urea and (b) melamine

Amine-formaldehyde (methylene glycol) condensation reactions occur in successive steps leading to the formation of methylol (—CH_2OH) groups or methylene bridges (—CH_2—), as in phenol-formaldehyde condensations:

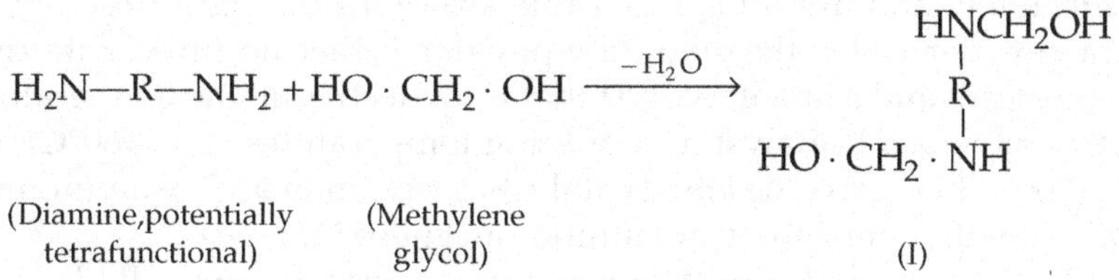

(3.17)

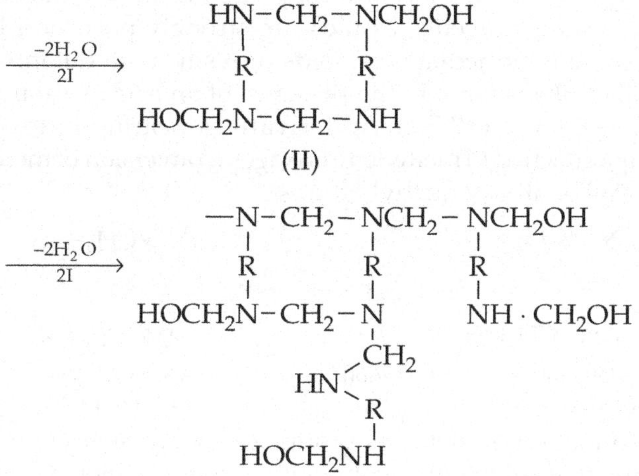

(III) (3.18)

For urea, R stands for C = O. Potentially, urea is tetrafunctional (presence of four N-hydrogens in each urea molecule) and melamine is hexafunctional. So, in each case, the successive step-growth condensation reactions lead to the progressive formation of more extended and branched structures with growing complexity, ultimately yielding highly cross-linked, dense network polymer structures.

For u-f resin making, urea to formaldehyde mole ratio of 1 : (1.5 or 2) is employed and 5% boric acid is used, maintaining a mildly alkaline (pH = 8–9) condition at 35–45°C initially and thereafter at 100°C for 2 h using Na_2CO_3 or NaOH or NH_4OH as catalyst. The initial resinoids, both monomethylol urea and dimethylol urea, soluble in water can be isolated in the pure form. They continue to react further engaging more formaldehyde and urea and engaging the growing methylolated step-growth species to form progressively big, complex, branched and mildly cross-linked products still maintaining water solubility. The resin in solution may be spray dried and stored for future use; the solution may be straight away used for plywood lamination in the presence of a latent acidic catalyst; or for making moulding powders using wood pulp with appropriate additives as filler, heated at 70–75°C for 2–3 h to allow impregnation and further growth in size of the resin molecules, which still retains fusibility but loses water solubility. Further heating of the impregnated pulp at 100–110°C nearly drives out much of the residual water; the initial relatively low molecular weight A-stage resin advances to the B-stage, characterized by higher molecular size and complexity. The resin-filler lumps harden effectively and are milled to a fine size with the right dose of pigments. The powder, called the moulding powder, is then admixed with right dose of a lubricant and a latent catalyst that is ineffective at low temperatures but effective as an acid catalyst at moulding temperatures (150–160°C) and pressures (1000–3000 psi). The latent catalysts in common use are ammonium sulfamate, trimethyl phosphate or ammonium chloride.

Urea-aldehyde resinoids and low-stage urea resins are not soluble in organic surface coating solvents, as they are strongly polar and hydrophilic. Condensing these intermediate resinoids or resins with butanol or a higher aliphatic alcohol like octanol in the presence of an acid as catalyst produces butylated or octylated (alkylated) derivatives bearing butoxy or related (alkoxy) groups attached to methylene linkages (conversion of methylol groups to butoxymethyl or alkoxy methyl groups):

$$\begin{array}{c} -\mathrm{N}-\mathrm{CH_2}- \\ | \\ \mathrm{C{=}O} \\ | \\ \mathrm{HN\cdot CH_2OH} \end{array} \quad + \mathrm{ROH} \xrightarrow{-\mathrm{H_2O}} \quad \begin{array}{c} -\mathrm{N}-\mathrm{CH_2}- \\ | \\ \mathrm{C{=}O} \\ | \\ \mathrm{HN\cdot CH_2OR} \end{array} \qquad (3.19)$$

(Methylol urea unit) (Alkanol) (Alkoxy methyl urea unit)

The transformation of methylol groups to alkoxy methyl groups renders the resin derivative soluble in weakly polar solvents; blocking the methylol groups renders the resin derivative less prone to cross linking, while allowing limited chain extension through interaction between free $—CH_2OH$ groups and residual $—NH_2$ or $>NH$ groups, thereby rendering the relevant product extensively soluble in surface coating solvents.

Melamine-formaldehyde (m-f) resins prepared with a melamine formaldehyde mole ratio of 1 : (2.5–3.5) are likewise prepared and used to produce moulded and laminated objects of far superior properties covering strength, hardness and gloss. Urea and melamine resins in water-soluble forms are useful as crush-proof, water-proof textile and paper finishing or sizing agents and the finish is made permanent and wash-resistant by heat setting. Much like methylol urea, methylol melamine may be transformed to alkoxy-methyl derivates to render the resin organo soluble and suitable for use in surface coating formulations. Blends of such alcohol-soluble urea or melamine-aldehyde resins and a range of compatible polyester (alkyd) resins produce enamels of outstanding merit, combining the best properties of each; the polyester acts as a plasticizer for the hard, glossy urea or melamine (i.e. amino) resins, while the respective amino resins act as hardeners for the polyester (alkyd) resins. The blend renders the coating enamel outstanding in hardness, scratch resistance, toughness, gloss, colour and colour stability, adhesion and outdoor durability. They are widely applied to coat automobile bodies, refrigerators and other consumer durables such as ovens and stoves, washing machines, etc. Baking or stoving is required to obtain full cure and high levels of property development. They are commonly referred to as stoving enamels.

3.4.11 Epoxy Resins

Epoxy or epoxide resins or polymers are produced by condensation or step-growth reactions of specified epoxy compounds as ingredients in the monomeric reactant mixtures. The most common epoxy resins are polycondensation products of epichlorohydrin and bisphenol A. Epichlorohydrin, the more expensive of the two, is derived from the chlorination of propylene via the formation of alkyl chloride and subsequently of 1, 3-dichloro, 2-hydroxy propane as intermediates. Alkali treatment of the latter leads to the formation of epichlorohydrin.

$$ClCH_2\text{—}CH(OH)\text{—}CH_2Cl \xrightarrow[-HCl]{NaOH} \underbrace{CH_2\text{—}CH}_{\setminus\, O\, /}\text{—}CH_2Cl \qquad (3.20)$$

The industrially-produced liquid/syrupy resins are essentially diglycidyl ether of bisphenol A (DGEBA) mixed with small proportions of some high molecular weight polymeric/oligomeric species. Two or three moles of epichlorohydrin are used for one mole of bisphenol A to obtain a higher yield of the diglycidyl ether. Commercial resins are available in the molecular weight range of 340–400, and very rarely, their molecular weight exceeds 3000–4000. The reactant mixture in the prescribed mole ratio is heated under stirring conditions to about 110°C under nitrogen atmosphere while a 30% solution of NaOH is added dropwise continuously to neutralize the liberated HCl. After the reaction, the organic layer is removed, dried and fractionally distilled under vacuum. A typical epoxy resin structure is shown in Fig 3.9.

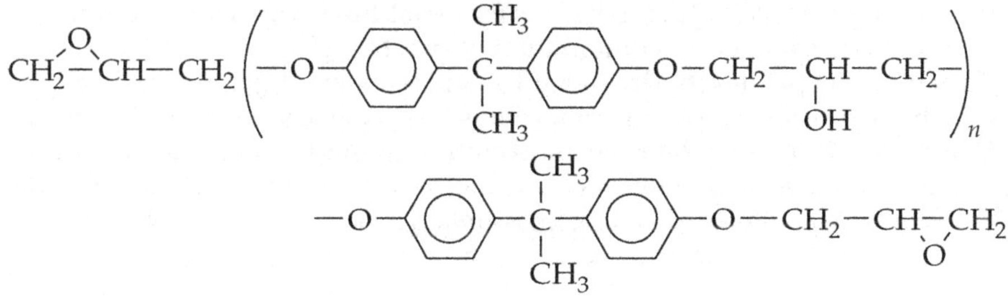

FIGURE 3.9 A typical epoxy resin molecule

Epoxy resins can be cross-linked utilising the epoxy and hydroxyl groups. The curing/cross-linking agents are known as hardeners/hardening agents. Mono-, di- and poly-amines, diacids and acid anhydrides are commonly used as hardeners.

$$\sim\sim CH_2\text{—}\overbrace{CH\text{—}CH_2}^{O} + RNH_2 + \overbrace{CH_2\text{—}CH}^{O}\text{—}CH_2\sim\sim$$

$$\rightarrow \sim\sim CH_2\text{—}\overset{OH}{\overset{|}{C}H}\text{—}CH_2\text{—}\overset{R}{\overset{|}{N}}\text{—}CH_2\text{—}\overset{OH}{\overset{|}{C}H}\text{—}CH_2\sim\sim \qquad (3.21)$$

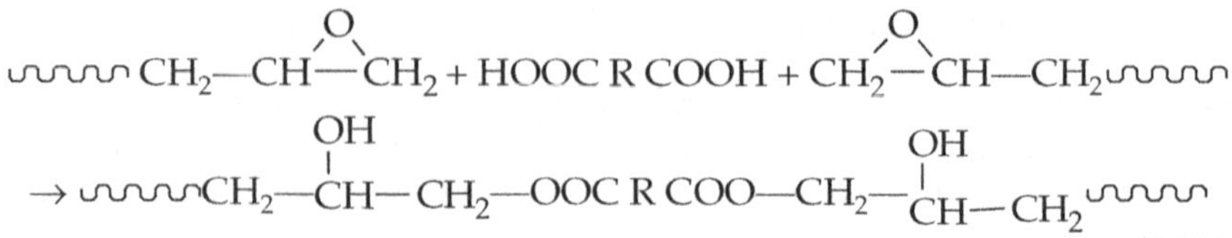

(3.22)

The reactions (3.21) and (3.22) are, in fact, linear chain extension reactions. Using diamines and polyamines with more than two active hydrogens in the molecule leads to three-dimensional networks. Using diisocyanates causes efficient cross-linking, forming what are known as epoxy-urethane linkages. Acid anhydrides react with the —OH groups initially present or formed by the opening up of the epoxide (oxirane) ring and effect cross-linking. Low molecular weight phenolic (novolac) and urea resins bearing methylol groups are also effective cross-linkers. Phenolic —OH or urea —NH_2 may also actively bring about cross-linking induced by diisocyanates. Amines are fast curing even at room temperatures and they impart good chemical resistance, even though they may cause toxic effects on the skin. Anhydrides are less toxic and yield cured products of good thermal stability; they are more effective at elevated temperatures.

Epoxy resins are suitable for moulding as well as lamination. They are especially useful as high strength adhesives and cementing agents in the building industry and road surfacing, in surface coating and in the potting and encapsulation of electronic components. Epoxy resins are characterized by excellent adhesion, very good cohesion, low creep, low shrinkage on cure and absence of volatiles. Different grades of epoxy resins are highly rated as adhesive, bonding and cementing materials for metal-to-metal, metal-to-ceramics, metal-to-plastics and other combination of substrates.

Reactions of epichlorohydrin, DGEBA and relevant oligomers with acrylic or methacrylic acid form intermediates called *epoxy-acrylates* or *epoxy-methacrylates* which may later be allowed to polymerize by means free radicals in the presence of selected acrylate, methacrylate, vinyl and related compounds to form chain-extended or cross-linked products of high cohesion and adhesive strength, toughness, clarity, colourability and durability.

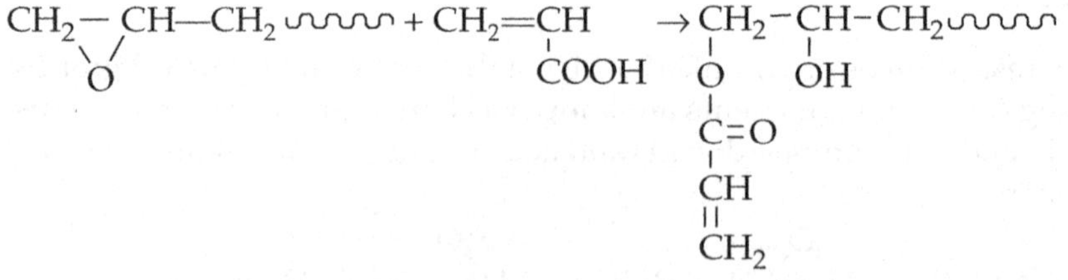

(3.23)

(Epoxy oligomer) (Acrylic acid) (Epoxy acrylate)

3.4.12 Vinyl Polymers

(a) Poly (Vinyl Acetate), PVAc

Poly (Vinyl Acetate), or PVAc, is widely used as an adhesive, particularly in emulsion form. A homopolymer of vinyl acetate or its copolymer prepared with alkyl acrylates (2-ethyl hexyl acylate), acrylic/methacrylic acid, N-mthylol acrylamide etc. as comonomers are used. Resin/Polymer emulsions of common use are available with 55%–60% solid content. A protective colloid (such as poly (vinyl alcohol), hydroxyethyl cellulose, starch or dextrin) is used to the extent of 3%–5%; emulsion polymerization is conducted using aqueous redox initiator systems such as $S_2O_8^{=}$ —$HSO_3^{=}$, $S_2O_8^{=}$—$S_2O_4^{=}$, or $S_2O_8^{=}$ — $S_2O_3^{=}$ combinations. Water-soluble phenolics, borax, ammonium rosinate, methyl cellulose and selected fillers may be used in the PVAc emulsion adhesive formulations to improve adhesion, remoistenable character, tack and body.

The largest adhesive application of PVAc emulsions is in bonding paper, particularly in packaging. The emulsions are used in book binding, carton-sealing, box-board manufacture and bag-seaming. A thin base coat of this is applied while making many polymer-coated fabrics. The PVAc emulsion adhesive is the most favoured adhesive in the furniture industry, where it has nearly completely replaced animal glues.

(b) Poly (Vinyl Alcohol), PVA

Poly (Vinyl Alcohol), PVA is derived from PVAc by acid or better alkaline hydrolysis of PVAc's ester groups. The degree of hydrolysis for useful products may vary between 75%–99.9%. Glycol, diethylene glycol or poly (ethylene glycols) may serve as plasticizers since dry PVA film is usually hard and brittle. 80%–90% hydrolysis of PVAc renders the derived PVA readily soluble in cold water and for hydrolysis >90%, the derived PVA is characteristically soluble in warm/hot water. Phenolic and amino resins may be used as agents for cross-linking and agents for improving the water resistance of the dry adhesive layer or the glueline. Fine clays, pigments and water-soluble polymers such as dextrin, starch and low-viscosity, water-soluble cellulosics are useful additives or modifying agents. 10%–20% of PVA in aqueous solution added to PVAc emulsion may dry up to give a remoistenable adhesive layer suitable for stamps and stickers. One major use of PVA as an adhesive or additive is as a component of PVAc emulsion adhesive.

(c) Poly (Vinyl Acetal), PVActal

This adhesive resin is obtained by reacting PVA with an aldehyde such as (a) formaldehyde, yielding poly (vinyl formal) (b) acetaldehyde, yielding poly (vinyl acetal) or (c) butyraldehyde, yielding poly (vinyl butyral) selected mixture of these aldehydes. The reaction of PVActal formation from PVA shown in Eq. 3.24 indicates the presence of minor proportions of residual acetate (—OAC) groups and lone, unpaired —OH groups along with the desired newly-formed acetal structural units in six-membered ring form in the structure of the actual PVActal molecules.

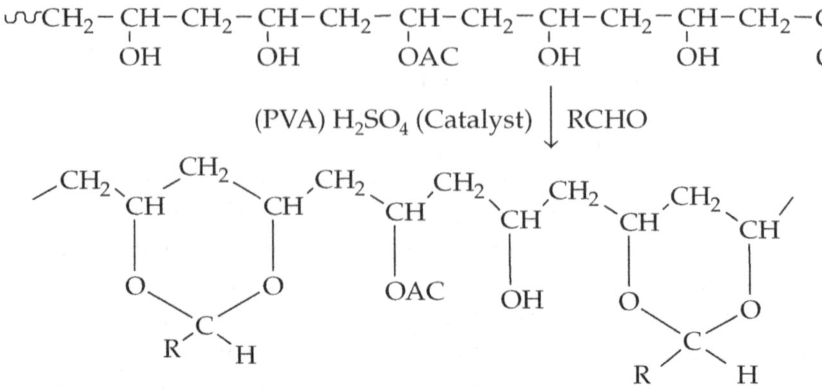

(3.24)

(PVActal)

Both partially hydrolyzed (15%–25% residual acetate groups) and nearly fully hydrolyzed (0.1%–5% residual acetate groups) PVA are converted to acetal forms using H_2SO_4 as the catalyst. When the reaction medium is water (starting with PVA in aqueous solution) for butyral formation, the initial homogeneous system finally turns heterogeneous as the reaction progresses; the butyral product formed, being insoluble in water, starts precipitating. If the initial medium is a water-ethanol mixture, the initial PVA/butyraldehyde system appears inhomogeneous, but with the progress of butyral formation, the system turns homogeneous in the end, because of butyral's solubility in the water-ethanol mixture. The separated butyral is completely precipitated by adding more water under stirring conditions, washed and dried.

The properties of the PVActal resins depend on the molecular weight, relative number and distribution of acetal groups, the residual acetate and the hydroxyl groups. When duly plasticized using dibutyl sebacate or triethylene glycol di-2-ethyl butyrate (40–45 parts per 100 parts of the resin), poly (vinyl acetal) or more appropriately, poly (vinyl acetal-butyral) finds extensive application as a clear, transparent interlayer in the making of automotive and related safety glass. Relevant resins also go into making printing inks and cements for binding metals together. Poly (vinyl formal)-phenolic resinoid hybrids, often called vinyl-phenolics, are rated very good to excellent as structural adhesives for metals.

Poly (vinyl methyl ether), prepared by aqueous polymerization of vinyl methyl ether using aqueous initiators, is water soluble and it finds some use in combination with some other water-soluble resins/polymers as a good adhesive.

3.4.13 Cellulosics

Cellulose is a natural fibre found in almost pure form in cotton and in the form of lignocellulose admixed with hemicelluloses in wood, bamboo, jute and other bast fibres like straw, grass, etc. Technologies have been developed to derive cellulose from most such renewable lignocellulosic resources.

Cellulose is basically a linear polymer, even though it defies solubility in water and most or all common solvents, and is infusible too, due to the existence of extensive intermolecular hydrogen bonds along the length of its chain molecules.

However, appropriately esterified (acetylated, nitrated, etc.) or etherified (methyl, ethyl or carboxymethyl) forms of cellulose are all soluble in appropriate solvents and are also fusible.[5–7] Some or much of the three —OH groups of the cellulose's, glucose units get lost and transformed to new functional groups (—$OCOCH_3$, —ONO_2, —OCH_3, —OC_2H_5, —OCH_2COOH, etc.) that are incapable of taking part in the establishment of H-bonding, thereby rendering the derived cellulose chains soluble and fusible.

Cellulose esters find widespread applications as moulded plastics, artificial fibres, films, sheets and as base materials for coatings and lamination.[5,6,7]

(a) Cellulose Nitrate

Commercial cellulose nitrates meant for plastics and surface coating applications have nitrogen content in the range of 10.9%–12% corresponding to degrees of substitution (DS) 1.8–2.3. Nitrates with an N_2-content of ≥ 12.5% (maximum attainable N_2-content for theoretical DS of 3.0 being 14.14%) are used as explosives.

The coating (lacquer)-grade cellulose nitrates are soluble in esters, ketones, ether-alcohol mixtures and glycol ethers (cellosolves). They are compatible with many other resins. Cellulose nitrate with 10.9%–11.2% N_2-content finds use in flexographic inks, lacquer coating for paper, foils and other flexible substrates.

(b) Cellulose (organic) Esters

The organic acid esters of cellulose, viz. the lone ester cellulose acetate and the mixed esters, cellulose acetate-propionate and cellulose acetate butyrate, when appropriately plasticized, are useful as tough, transparent plastics (sheets, films widely used for photography and wrapping), light envelopes, bags and for packaging. The water absorption of mixed esters is lower than that of the lone acetate ester; as such, the mixed ester, cellulose acetate-propionate has some application in surface coatings and the secondary cellulose acetate (diacetate) is useful as a man-made fibre—known in the trade as (acetate) *rayon*.

(c) Cellulose Ethers

The most important of the cellulose ethers is ethyl cellulose. The commercial material is prepared by reacting cellulose with an alkyl halide (ethyl chloride) in the presence of NaOH (swelling agent) at 60°C for several hours; the temperature is subsequently raised to boiling point and the ethyl alcohol and ethyl ether formed as byproducts are removed by distillation. The product, ethyl cellulose, is steeped in hot water, centrifuged, washed and dried.

$$\sim\sim\text{Cellulose}—\text{OH} + \text{RCl} + \text{NaOH} \rightarrow$$
$$\sim\sim—\text{Cellulose}—\text{OR} + \text{NaCl} + H_2O \qquad (3.25)$$

The fully substituted product (DS = 3.0) exhibits an ethoxyl content of 54.88%. With DS value of 0.8%–1.2%, the product ethyl cellulose is water-soluble. The highly or fully substituted product, however, is soluble in non-polar solvents (aromatic hydrocarbons). Ethyl cellulose is used in hot-melt, strippable coating formulations and in surface coating formulations (paints, varnishes and lacquers). Water-soluble grades are useful as textile finishing and paper sizing materials.

Methyl cellulose (DS 1.6–2.0) is water soluble; it is employed as a thickening agent and as an emulsifying and paper sizing agent, and to a limited extent, in leather tanning and ceramic processing. Hydroxyethyl cellulose prepared by the reaction of alkali cellulose with ethylene oxide also finds applications similar to methyl cellulose. Carboxymethyl cellulose (CMC), (~— Cellulose —O—CH_2COOH) is formed by the reaction of soda cellulose with a sodium salt of monochloro acetic acid and subsequent acidification. Na-CMC of DS in the range of 0.5–1.0 is water soluble and commercially useful as an additive in synthetic detergents (soil-suspending agent) and as a stabilizer or viscosity modifier in aqueous suspensions and emulsions, besides being useful as a textile and paper sizing agent and as a thickening agent in food, pharmaceuticals and cosmetics.

3.4.14 Polymer/Resin Additives

In general, polymers/resins of any class, meant for application as plastics, rubbers, adhesives, coatings, lamination and encapsulants, are seldom used in the pure form. In most cases, they are mixed or compounded with a host of additives to improve application rheology and processibility, stability, durability and service life, property range and performance-cost benefits.[6–8] Functionally, the additives may be categorized under the heads given in Table 3.4; the figures in the right column indicate parts by weight of the respective ingredients for 100 parts of the base matrix resin or polymer used. However, it is not absolutely necessary to use all the ingredients in all usable formulations.

TABLE 3.4 Additives for polymers (rubbers, resins and plastics)[6, 8]

Sl. No.	*Ingredients*	*Parts by weight*
1.	Matrix resin/polymer	100
2.	Filler	20–60
3.	Plasticizer	10–60
4.	Stabilizers:	
	(a) Antioxidant	0.5–5
	(b) Antiozonant	0.5–5
	(c) UV-absorber	0.5–2
	(d) Thermal stabilizer	2–15
	(e) Stabilizer against microorganism attack	1–3
5.	Lubricant	1–2
6.	Flow promoter/Tackifier	2–15
7.	Accelerator	1–15
8.	Cross-linking agent	1–12

(Contd.)

Table 3.4 (*Contd.*)

9.	Colouring agent	2–6
10.	Antistatic agent	1–3
11.	Flame retardant	2–10
12.	Others (retarder, solvent, dispersant, compatibilizer, coupling agent, activator, blowing agent, thickening agent etc.)	1–5 and 20–400*

*For solvents/dispersants/diluents

The technologist utilizes the tack property of the matrix resin/polymer material (melt tack or solution tack) to uniformly distribute the additive items into the resin under appropriate shear action on the melt, or by stirring the solution. The resin picks up the additives and raises its storage stability, processibility and performance; it becomes largely cost-effective, attractive and durable during storage, processing and use. When used in solution or dispersion form, as for adhesives and surface coatings, the solvent or the dispersion medium is to be driven out, often on heating or allowing air drying (sometimes aided by the application of vacuum or air flow) before the matrix polymer is set to the desired solid interlayer or surface film. Blending is commonly accomplished by using a dry blender (rotating drum/ribbon blender),then a hot open (two roll) mill, followed by cooling and grinding. Alternatively a pot fitted with a stirrer and optionally a heating device for handling the resins/polymers in the presence of a solvent, plasticizer or dispersion medium to turn them into solution, paste or dispersion may be used.

References

1. Tanford, C, *Physical Chemistry of Macromolecules,* John Wiley and Sons, New York, 1967.
2. Robert, J D and M C Caserio, *Basic Principles of Organic Chemistry,* W A Benjamin, Inc., New York, 1965.
3. *Shellac,* Angelo Bros., Calcutta, 1965.
4. Rangaswami, M and H K Sen, *A Handbook of Shellac Analysis, Indian Lac. Res. Inst.,* Namkum, Ranchi 1952.
5. Ghosh, P, *Fibre Science and Technology,* Tata McGraw-Hill, New Delhi, 2004.
6. Ghosh, P, *Polymer Science and Technology—Plastics, Rubbers, Blends and Composites,* 2/E, Tata McGraw-Hill, New Delhi, 2002.
7. Schmidt, A X and C A Marlies, *Principles of High Polymer Theory and Practice,* McGraw-Hill, New York, 1948.
8. Ghosh, P, *Polymer Vijnan O Manab Samaj* (Bengali), ARGHYA, ISEC and PSMPAF, Kolkata, 2005.

Chapter

4

Adhesives: Principles and Practice

4.1 Adhesion and Bonding

Adhesion is a process by which two similar or dissimilar *adherend* surfaces are partly or wholly held together[1,2] in close contact by:

(i) *surface attachment* or *interfacial forces of attraction* consequent to interactions of molecules, atoms or ions in the two (adhesive–adherend) surfaces facing each other, or by

(ii) *mechanical interlocking*.

The adhesion process is aided, in most cases, by the presence of a thin interlayer of an organic resin or polymer, natural or synthetic, manipulated by spreading its solution or melt and allowing the spread-out interlayer to display cohesion by the interplay of solution or melt tack. The interlayer is finally allowed to set and harden by solvent evaporation and/or cooling for strength. This concept is not to be conventionally applied to metal solders, even though one is inclined to view soldering as an adhesion process in every sense. The two bodies held together by adhesion are called *adherends* or *substrates*, even though the latter term may be broadly used for other bodies having different roles or functions. The term "bonding" with respect to adhesives is meant to denote the process of joining or fixing of surfaces together by a process of adhesion, i.e. by adhesive action. The adhesive interlayer, together with the adherend-adhesive interfaces on the two sides, is commonly referred to as the *glue-line*. Figure 4.1 gives a description of an adhesive joint.

Rheological characteristics and demands favour specified polymers and resins as good candidates for use as adhesives. For proper spreading and development of tack, adhesives are almost invariably applied in liquid form or are necessarily liquified at some stage in the bonding process and allowed to

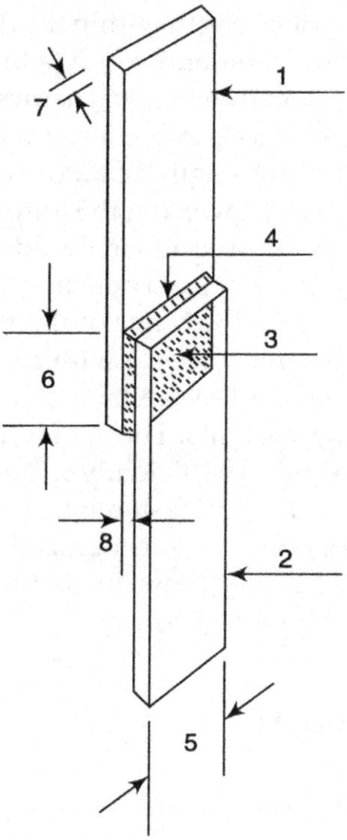

FIGURE 4.1 An adhesive joint: (1) first substrate; (2) second substrate; (3) interface/bond surface; (4) flash; (5) bond width; (6) overlap; (7) substrate thickness, and (8) bond gap/glue line thickness

dry and set to infuse permanence and strength. The setting or hardening may involve:

(a) just cooling to solidity or simple evaporation of the solvent
(b) absorption of the solvent by porous adherends allowing solvent/volatile escape through the pores even after joint formation
(c) liquid-solid or sol-gel transformations
(d) reactive processing involving chain extension, polymerization and/or cross-linking or a combination of some or all of these, leading to increase in viscosity, setting of the monomer/oligomer/prepolymer system

Setting enhances the joint strength, which in many cases be higher than the strength of the adherends themselves. While adhesives set to thin films that join two surfaces firmly, surface coatings simply provide thin films that cover and adhere to one surface to offer protection, and in many cases work as decoration, too.

4.2 NATURE OF ADHESIVE JOINTS

For a strong, durable adhesive joint, it is necessary to establish adhesion at the interfaces between the adhesive and the adherends and to ensure the development of good cohesion or strength within the dried-up adhesive layer itself. Let us take the example of water and wood (a lingocellulosic material). The similarity in polar character and the chemical natures of water and cellulose is manifested in good wetting; however, spreading of water between two blocks of wood offers a joint of negligible strength. But if the water layer is converted to ice, of high cohesive strength, by lowering the temperature, the joint turns so strong that when put to test, failure occurs in the adherend wood and not at the interface or in the adhesive layer of ice. There may be five possible regions of joint failure, viz. at (1) one or (2) the other of the two adhesive-adherend interfaces, working against adhesive forces or within (3) one or (4) the other of the adherend strips and (5) in the adhesive layer itself, working against the cohesive forces in the respective systems. The actual rupture will occur in the weakest part of the assembly. When a sufficiently strong and durable adhesive joint is developed to replace mechanical fastening, say of metal strips or parts (using screws, rivets, nuts and bolts), it is recognized as a major engineering innovation and practice. Adhesive joints in metals may help eliminate bimetallic contact and the associated corrosion.

4.3 Adhesion Mechanism

4.3.1 General Considerations

The science of adhesion may be viewed under two categories:

(1) one dealing with the chemistry and physics of surfaces and interfaces, and

(2) relating to the mechanics of fractures in adhesive joints

The first, commonly referred to as *specific adhesion*, concerns bond formation, i.e. wetting, adsorption, chemical interlinking, etc., while predicting the magnitude of the intrinsic adhesive strength across the interface. The latter, *mechanical adhesion*, concerns the mechanical analysis of the strength of adhesive joints and the development of test methods to measure the joint strength.

The interaction, i.e. adhesion between two different types of materials across an interface, may involve chemical or physical bonding. Chemical bonding involves direct interlinking between molecules of the two materials, viz. the adhesive and the adherend, by covalent and ionic bonds. Physical bonding arises from mechanical interlocking or from forces of physical adsorption between the molecules of the adhesive-adherend pair or else by the penetration of adhesive molecules into the adherend substrate by diffusion.

4.3.2 Mechanical Interlocking

This mode of adhesion arises when the adherend surface has pores into which, or protrusions/projections around which, the spread out adhesive may flow and solidify to provide mechanical anchorage. Even if the intrinsic chemical interaction between the molecules of the adhesive and the adherend is low, high joint strength can be achieved in this manner. *Mechanical interlocking* has an important role to play while bonding pieces of substrates such as wood, paper, textile and specific clays, due to their finely divided nature and porous character. In many instances, the adherend surface (such as the surfaces of metals, plastics and rubbers) needs to be etched before adhesive application and bonding so as to enable the spreadable/flowable adhesive mass to penetrate and get locked into the adherend substrate. For some substrates with very smooth and polished surfaces, like glass and some polished metals, this mechanism is of little consequence.

Joint strength attributed to mechanical interlocking is often traceable to other effects. Roughening of an adherend results in improved bonding by increasing the total available surface area or by exposing fresh clean surfaces with strong unbalanced valence forces, rather than by affording any purely mechanical interlock. It is common experience that sanding wood surfaces with fine sand paper just before glue application results in a stronger bond than that obtained using glazed surfaces or surfaces that have been roughly sanded.

4.3.3 Interdiffusion, Adsorption and Surface Reaction

This mechanism involves molecular mobility. The liquid adhesive may dissolve and diffuse into the intermolecular interstices or voids of the substrate material. The degree of diffusion is contingent upon the different types of intermolecular affinity at the contact layer. The molecules of the adhesive base material are usually polymeric and exhibit only limited compatibility with most substrate molecules; in that event, the interdiffused layer is usually very thin (< 10 nm). For substantially high adhesive-adherend compatibility, however, the interdiffused layer may turn out to be much thicker (10 μm).

Adsorption is the process by which a molecule of an adhesive is attracted to a suitable site on a solid surface and rests there. The greater the polarity of the relevant molecules, the greater the molecular attraction between them. Molecules of the adhesive and adherend may also possess specific chemical characters (polarity, sites for H-bonding, ionic groups covering acidic and basic structural units) that enhance adhesion and chemisorption. The adsorption and surface reaction mode of adhesion is more relevant for organic adhesives on high-density substrates such as metals and glazed ceramics.

4.3.4 Electrostatic Attraction

In a combination of an adhesive and adherend with very different electronic bond structures, electrostatic forces develop at their interfaces. These forces are commonly attributed to the transfer of electrons across the interface, thus creating positive and negative charges that attract one another; an attracting electrical double layer is thereby created. Serious reservations, however, exist in the minds of one school of scientists about the magnitude of such electrostatic attractions, compared to Van der Waals interactions, and that the electrical double layer apparently makes a small or insignificant contribution to adhesion,[3] even though a different school of scientists, however, maintains that the role of electrostatic forces in adhesion is dominant.[4]

4.3.5 Wetting and Setting

Since the function of an adhesive is to join two solids together, it must be able to make intimate contact with each surface and spread freely upon them, i.e. it must be capable of *wetting* the surfaces. The adhesive must be applied in the liquid state and with enough pressure to make it flow into the small pores and crevices of the solid surface. Furthermore, it must displace or dissolve adsorbed impurities such as water, air or oxygen; otherwise a weak boundary layer having some contactless pockets will be formed. It is only then that the adhesive layer is capable of setting to a thin, tough, sticking, cohesive solid layer. Also, in order to minimize the internal stresses in the adhesive joint, there should not be a large change in the volume, i.e. shrinkage, of the adhesive on solidification, and the thermal expansion coefficients of the adherends and adhesive should preferably be similar.

The setting of adhesives occurs in three different ways: by cooling, by solvent evaporation/removal, and/or by chemical reaction (polymerization of monomers, oligomers and prepolymers or cross-linking). Solvent-based adhesives suffer the greatest shrinkage during the setting process.

The degree to which a liquid wets a solid is measured by the contact angle θ, as in Fig. 4.2. When $\theta = 0$, the liquid spreads freely over the surface and is said to completely wet it. Complete wetting occurs when the molecular attraction between the liquid and solid molecules is greater than that between like liquid molecules.[5]

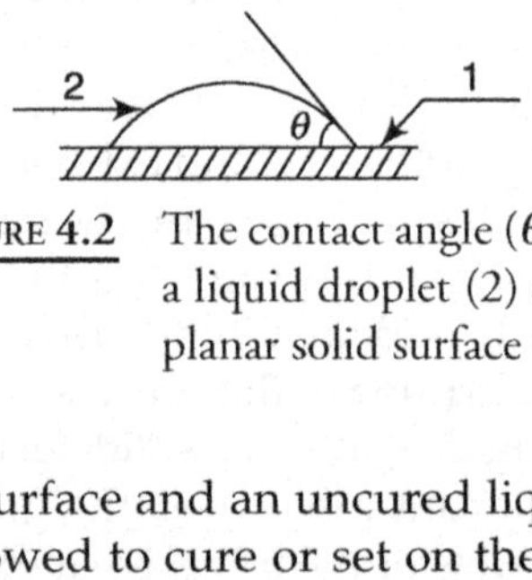

FIGURE 4.2 The contact angle (θ) of a liquid droplet (2) on a planar solid surface (1)

The adhesion between a polyethylene (PE) surface and an uncured liquid epoxy is quite low even when the epoxy is allowed to cure or set on the PE substrate. However, if the PE is melted and then spread on a cured epoxy substrate, the adhesion is much stronger. Liquid epoxy, having high surface tension (γ_l), will not wet a low surface-energy solid such as PE (γ_s) because, in this case, $\gamma_l >> \gamma_s$. However, for the second case, wetting is substantially favoured as the liquid PE is of lower surface energy (γ_l) than the solid cured epoxy (γ_s), i.e. for $\gamma_l << \gamma_s$. The case also assumes significance because molten (liquefied) PE has a high viscosity (> 10^3 Pa.s) compared to the viscosity levels of liquid epoxies commonly used (~ 1 Pa.s). The wettability of the solid surface by the spreadable liquid is, however, more important than the viscosity of the spreadable liquid.

4.4 Surface Treatment and Topography

To obtain a strong, and durable adhesive joint, the surfaces of adherends are often cleaned and/or treated before bonding. In general, these treatments alter the surface region in one or more ways: removal of a weak surface layer; desired change in surface topography; change in the chemical nature of the surface; or modification of the physical nature and structure of the surface, some of which are illustrated next, considering polyolefins or metals as adherends.

4.4.1 Polyolefins

Nearly all adhesives have surface tensions too high to allow them to wet surfaces of solids having low surface energies, e.g. polyolefins and polyfluorocarbons. Several methods have been devised and are in practice for modifying the surfaces of these materials such that they may be strongly bonded with conventional adhesives and printed with polar resin (alkyd)-based inks.

Controlled flaming, plasma treatment or cross-linking by activated species of inert gas (CASING), corona discharge and acid etching (chromic acid is commonly used to generate polar —OH, —COOH, $>C{=}O$, —SO_3H groups) are conventional techniques used for surface treatment and the modification of polyolefin and related polymers to achieve improved, durable adhesion and easy printability. Both corona discharge and acid etching lead to the roughening and oxidation of polyolefin and polyfluorocarbon surfaces, thereby facilitating higher joint strength. Increasing the etch time with chromic acid not only creates a high degree of surface oxidation, it also infuses a higher depth of etching and oxidation, facilitating higher adhesion, easy printability and higher adhesive joint strength.

Studies have shown that CASING causes cross-linking mostly in the surface layers. When a polymer is cooled from the melt, a low molecular weight fraction is pushed to the surface, before growing crystallites nucleated in the interior of such polymers as polyethylene (PE). This may result in a surface region that may act as a weak boundary layer in an adhesive joint made from hot melts. Cross-linking the adhesive near the substrate surface region enhances the strength of the adhesive layer that anchors on the substrate surface; in the process, this leads to considerable reinforcement of the interface and hence enhancement of the joint strength as well. The fracture of an adhesive bond between an untreated PE substrate and an epoxy adhesive occurs clearly at the PE-epoxy interface, but not within the PE surface layer. A good feature of epoxy resins is that they shrink only by $\leq 3\%$ on setting.

The PE (or polyolefin) material is given a *corona discharge* normally in air under normal pressure; $>C{=}O$, —COOH and —OH groups and C=C unsaturations develop on PE or other polyolefin surfaces as a result.[6,7] Besides this, there is some or even a substantial degree of roughening in the surface, indicating that corona discharge also causes measurable surface degradation and some degree of surface material removal in a non-uniform way. The amorphous regions suffer more effectively from corona discharge. The much enhanced bonding to a corona discharged PE surface is consequent to both development of higher surface roughness and to a notable increase in the critical surface tension or surface energy.

Also, minor extents of surface grafting (on pre-treating the PE surface with UV light of γ rays in the presence of a relatively polar monomer (methyl acrylate or vinyl acetate) on PE and a related non-polar substrate leads to substantial improvement in bond strength using say, epoxy adhesive, as in Table 4.1.

TABLE 4.1 Effect of methyl acrylate graft copolymerization on polyethylene* on its attachment to epoxy adhesive**

Methyl acrylate on PE surface, mol %	*Peel strength N/m*
10	60
20	100
60	810
80	1610

*UV-induced grafting (pre-irradiation/polymerization under N_2)
**Polyamine cured

4.4.2 Metals

To prepare a metal surface for bonding (for application of an adhesive or surface coating), etching techniques may be followed to remove the existing oxide layer as well as any organic surface contaminants. Chemical etching may also end up removing some of the metal from the surface. The metal at the surface or near it may appear significantly different in physical structure from the bulk; however, this largely depends on the particular technique followed during processing and forming.

The surface treatment of aluminium (Al) has received much attention. Al-epoxy bonds are being extensively used in aerospace technology. To create an Al-surface layer that would form an adhesive joint of the highest possible durability and strength, the top oxide layer must be removed before spreading the adhesive formulation, which is commonly based on epoxy, phenolic, epoxy-phenolic, epoxy-acrylate, epoxy-urethane resins or selected blends thereof. An optimum degree of anodizing using an aqueous solution of phosphoric acid (phosphoric acid anodization, PAA) for several minutes, followed by rinsing and then drying in hot air is reported to be of proven advantage in this context.

Electropriming is another form of metal surface treatment through which an oxide layer and a simultaneous deposition of a polymer layer are achieved. A cleaned metal (adherend) surface is made the anode or cathode of an electrochemical cell containing an aqueous primer-resin suspension. Adhesive joints on such pretreated metal substrates are strong and durable. The beneficial role of using selected silane coupling agents in surface treatments has been discussed in Chapter 1. Silanes bearing vinyl, ethoxy, glycidoxy propyl and amine groups act as good coupling agents between adhesive resins and wood, glass, metal and ceramic surfaces during adhesive bond formation.

4.4.3 Other Points of Importance

Some pairs of adherends differ so much in chemical and structural properties that suitable adhesives are rarely found. It then becomes helpful to deliberately alter the surface composition of one of the adherends. This is accomplished by chemical reaction or applying a coating.

The bonding of steel and rubber in tyre construction is an example. For making tyres of good form and load-bearing capacity, the steel bead wire used in the tyre rim must be bonded strongly with rubber in the side wall and tread. To achieve the objective, the steel bead wire is given a thin coating of brass; the rubber in the tyre is then vulcanized under pressure in close contact with the brass-coated steel. The consequent rubber-steel wire adhesion becomes excellent due to the formation of primary-valence copper sulphide bonds between the copper in the brass coating and the sulphur distributed and dissolved in the rubber. This kind of technology approach is somewhat similar to phosphating and chromate priming in the surface coating of metals and alloys.

4.5 Polymer Characters for Adhesives Application

For the base polymer making the major part of an adhesive formulation, a broad molecular weight distribution may be better than a narrow one for some good reasons: smaller molecular weight fractions of the applied polymer-adhesive, characterized by higher mobility and penetrative capacity, promote wetting and specific adhesion. As the adhesive layer sets, the very large molecular weight fractions will precipitate in their own bulk or phase-separate and gel first. They will eventually contribute greatly to the development of cohesive strength in the adhesive layer. The substantially more mobile, smaller chain molecules will form a solution or blend of relatively low viscosity that will migrate toward the adherend surfaces to not only wet the interfaces, but also to cause the ready displacement of adsorbed air and other atmospheric gases and penetrate readily into the adherend mass through surface crevices and pores. It is thus easy to envision molecules of different sizes functioning with specificity and effectiveness at appropriate locations, which perhaps could not have occurred if the chain molecules were all of one size.

Polymers used as adhesives are either linear or branched thermoplasts, or space polymerizing and network-forming thermosets. The latter, in particular, possess the intrinsic characteristics of being solvent-free and applied to the joint as small, mobile molecules of low viscosity, high wetting and penetrative power and they are finally polymerized *in situ* to form cross-linked, network polymers of much higher rigidity, stiffness, toughness and solidity than the linear/branched thermoplastic polymers.

4.5.1 Volatile Release and Removal

Adhesives are most widely formulated as solutions in organic solvents, water or in aqueous dispersions. The release and removal of the volatile solvents and the dispersing water is of importance for high performing adhesive joints. For room temperature adhesive application, a critical control over volatile content at the point of closing of the assembly is required. If too much is present, excess volatile matter may get trapped in the closed assembly and weaken the final bond by overplasticization and/or void generation. If too little is present, the fusion or attachment of the two spread-out surfaces, wetting of the adherend and transfer of adhesive mass from a spread-out surface to a bare, unspread surface will be incomplete. The overall formulation, manipulation and control should match the required optimum time interval between spreading, volatile escape and closure. Overdrying before closure must be avoided; underdrying is less critical for porous adherends like wood, paper-board etc. than for impenetrable adherend substrates. For a pair of impenetrable adherends such as glass or metal, all volatile matter must be expelled after spreading; subsequent application of heat under positive pressure on closing the assembly would ensure the maximum joint strength.

4.5.2 Aqueous Adhesives

Water-soluble glue-lines are likely to swell, weaken and delaminate in contact with water or they may weaken substantially in moist weather. Glues of vegetable or animal origin may undergo microorganism or chemical attack when moist. Certain preservatives in low doses may guard against microorganism attack. Adhesives based on starch, dextrin, gum arabic, casein, gelatin and blood are to be used keeping these inherent possibilities and limitations in mind. However, they become more resistant and durable when applied in the presence of a little formaldehyde or formaldehyde-generating chemicals (hexamethylene tetramine, paraformaldehyde, etc.) as curatives. Ready availability at low cost, environment-friendly character and non-toxicity are the points in favour of using water as the solvent or dispersion medium for adhesives and paints. Many organic solvents evaporate more readily and rapidly than water. Organosoluble adhesives are less likely to be sensitive to atmospheric changes. For most solvent-based adhesives, a solids content of 15%–40% is common. The base polymer/resin in water-based adhesives include phenolics (resoles), amino (urea, melamine) resins, rubber lattices, poly (vinyl acetate) (emulsion form), poly (vinyl alcohol), poly (vinyl methyl ether), gum arabic, starch and dextrin, and protein (collagen, soybean and casein) glues, while organic solvent based adhesives include duly masticated, compounded and curable natural/synthetic rubbers,[8] acrylate/methacrylate polymers/copolymers, polyurethanes, dimer acid-based polyamides (also used as hot-melt adhesives), poly (vinyl acetals) and some plastisols.

Many adhesives in the market are available as emulsions. Here, polymers are synthesized in emulsion from the monomeric system. Polymer emulsions having a 30%–40% solids content offer much lower a viscosity than a corresponding system in solution. On the other hand, PUs, epoxies, silicones and some elastomers may be ground fine and dispersed by high shear in the presence of an emulsifier to obtain water-borne dispersion formulations. In such formulations, it is sometimes customary to disperse alongside a finely-divided, low melting, solid plasticizing resin to develop *delayed tack* adhesive; they develop enhanced tack on heating, when the plasticizing resin melts and in the process, the initially obtained solvent-dried, tack-free glue-line additionally functions as a hot-melt, tack-developing adhesive system, finally developing higher bond strength and showing better quality and efficiency, with a possibly thinner glue-line.

4.5.3 Bio-adhesives

All natural glues, animal/vegetable protein-based or polysaccharide-based, and some other glues/adhesives or membranes/gums that are bio-compatible and bio-degradable or bio-resistant are useful as bio-adhesives. Some are also useful as restoratives and cementing agents—such as for making gum and tooth impressions using heat-softened compositions in dentistry, and in burn or infected wound-healing and skin restoration purposes. They use polycarboxylates based on poly (acrylic acid) and selected metal oxide fillers and chitin/chitosan-based antibacterial cements/membranes often attached to skin sites, using absorbable bio-sutures (also see section 3.2) if required.

Bio-adhesion may be viewed as a state when two materials, one of which is of bio-origin, are held together for extended periods of time by interfacial forces. In medical and pharmaceutical contexts, bio-adhesion means sticking together synthetic and biological macromolecules to a biological tissue. The biological substrates are cells, bone, dentine or the mucus coating the surface of a tissue.

In healthcare, bio-adhesives are used as wound dressings, skin adhesives and denture cements. Moisture-activated bio-adhesives (alginates, gelatin, poly (acrylic acid) salts/derivatives are also being fabricated and used as essential components of medical devices, medical sensors and drug delivery systems. They readily attach, stick and conform to accessible sites of the body. Dried poly (acrylic acid) hydrogel films are readily formed for bio-adhesive application in the presence of aqueous fluids.

A range of quick-setting UV-curable bio-adhesives are given by copolymers of N-vinyl pyrrolidone and some other monomers such as vinyl succinimide, glycidyl acrylate, etc. Chitosan's properties permit it to rapidly clot blood that has prompted its approval for use in bandages. Its adhesive character makes chitosan an excellent candidate as a flocculating agent in the food and beverage industry.

The sticky exudates of frogs of the genus Notaden is reported to set into a tacky, elastic, bio-compatible solid that readily adheres to a wide range of materials such as glass, plastic, metal, wood, fluoro polymers and skin even when wet. The frog-glue bonding is nearly as good as bonding using PVA and cyanoacrylate glues.

4.6 Hot-melt Adhesive Systems

A thermoplastic polymer that becomes flowable, spreadable and tacky on heating, and solidifies on cooling after application in melt form, is commonly referred to as a hot-melt adhesive. In real hot-melt systems, the polymer/resin should have molecular weights on the lower side for adequate fluidity at the high application temperatures. Sometimes, low molecular-weight non-volatile additives (coumarone-indene resins or petroleum resins in the molecular weight range 2000–5000) are incorporated to further drive home the advantage. The adherend surfaces need to be pressed together while the thermoplastic polymer/resin adhesive is still in the melt/fluid condition. The more common and important hot-melt adhesives are:

(i) polyethylenes or other polyolefins and polyolefin blends
(ii) ethylene-vinyl acetate (EVA) copolymers and their hydrolyzed versions (EVOH), which offer additionally excellent barrier properties with respect to gas (oxygen) permeability
(iii) selected grades of copolyesters/copolyamides compounded with novolac resins and fillers, and also dimer acid-based polyamides popular as bonding agents and sealants in furniture, electronics and footwear industries
(iv) linear (nylon) polyamides, polyesters, polyurethanes, (PU), and
(v) high-styrene SBR resins, etc.

Multilayer polymer laminates suited for special packaging applications are manufactured by employing *coextrusion* technology. Thin films of the appropriate polymers that can be produced from an extruder may be laid between two non-meltable, heat-proof adherend substrates and the assembly is duly stuck together and pressed if necessary, to attain the relevant adhesive bond. Hot-melt, solvent-free adhesives are commanding an increasing market share on considerations of 100% solids left, handling of non-volatiles, quick operation, low cost and environmentally friendly features. Sometimes, the polymer to be used is spread in a powder form on flexible substrates, and then melt-pressed between rollers, as in the case of polyethylene-bonded papers and textiles. Filled EVA hot-melt adhesives are in wide use in the book binding and footwear industry, as heat-activated adhesive tapes (carpet seaming) and in carton sealing and furniture end veneering.

We often use *film-adhesives* as heat-activated bonding agents, applied much like hot-melt adhesives; they also resemble each other in the mechanism of bonding. Polymers as a preformed sheet or film offer advantages such as simple or rapid bonding; the availability of duly designed two-sided films permits the use of an adhesive of different character for each of the two (unlike) adherend surfaces to be bonded. Epoxies, phenolics, nitrile rubber, PVC-nitrile blends, polyamides, poly (vinyl butyrals) and ethylene-carboxylic acid-based ionomers are all available in sheet/film forms for adhesive jointing of different adherends, like or unlike. The films are usually laid on a release paper for storage/handling and convenient application and they can be preshaped/prefabricated to match the shape of the adherend substrate surface. Non-woven webs, woven fabrics or knitted nets from natural/synthetic fibres are likewise used as substrates for film coating in making bags, luggage, furniture seats, automative interiors footwear and related industries. Film adhesives may be used for bonding fabrics and foams, and films/sheets and foils of different kinds.

4.7 Polyurethane (PU) Adhesives

We may have one component PU-adhesives and two component type PU adhesives each of which is also recognized as reactive adhesive type much like the phenolics, urea resins, epoxies, etc. They may be applied from solutions, used in film form as an interlayer (hot-melt type), and in aqueous dispersions/emulsions. The reactive adhesives, able to further establish urethane or urea type bonds by step-growth reaction mechanism after application, are by far the most important.

For making PU-based reactive adhesives, the more important isocyanates used are toluene diisocyanate and diphenyl methane -4-4′-diisocyatate (MDI). The diisocyanate chosen is allowed to react with linear diols or branched polyols to produce the so called *prepolymers.* Depending on the —OH/—NCO ratio used, the prepolymer may be —NCO or —OH terminated. The oligomeric diol or polyol in use may be a polyether or a polyester and the relevant prepolymer obtained may be called polyether urethane or polyester urethane.

The isocyanate-terminated prepolymers are useful by themselves as a one-component adhesive system and are usually applied in bulk or in solution form. Curing readily occurs by utilizing moisture present in the air or in the adherends; the curetime would depend on the thickness of the glue-line. The moisture-cure liberates CO_2. With the high-NCO content in the prepolymer, desirable for the development of greater adhesive bond strength, the high amount of CO_2 liberated may make the finally dry adhesive layer porous. So, moisture-cured PU-adhesives are unsuitable as structural adhesives. They are commonly used in the packaging industry, joining films of PE, cellophane, paper, nylon, rayon and polyester.

Sometimes, polyisocyanates (not true PUs) are used by themselves to bond two adherend substrates. Elastomers may be coated with the solution of a polyisocyanate and vulcanized in contact with metals after drying. Very strong, solvent-resistant bonds are obtained. An elastomer may otherwise be compounded with a polyisocyanate and a cement is prepared by dissolving the compounded item in a dry aromatic solvent. The cement promotes the development of very high bond strength between elastomers, papers, fabrics, metals and ceramics.

One-component, heat-curing adhesives are based on phenol-blocked diisocyanates. At high temperatures, the blocking group (i.e. phenol, ketoxime or caprolactam) is released and removed, liberating the isocyanate groups for reaction to form urethane and related structures. Further, dispersions of insoluble polyhydroxy compounds in an isocyanate-terminated prepolymer may result in a product where reaction occurs only when the mixture is heated and a single-phase results.

Two-component PU adhesives utilize isocyanate-terminated prepolymers as one component for reaction with a polyol or polyamine as the second component. When the components are mixed in the desired proportions and applied, a short pot life, i.e. a fast cure is desirable. A faster cure is achieved by heating the bonded assembly, preferably while keeping the assembly under pressure and clamping. The two-component reactive urethane adhesives are used as structural adhesives in the construction and automotive sectors.

Aqueous dispersions of selected PUs can be obtained by using surfactants; a more important class of dispersions is made by incorporating ionic functions in the main chain. A reaction of the diisocyanate-terminated prepolymer with a diamine containing a carboxylic acid or sulfonic acid in a water-immiscible organic solvent yields an ionomer, which is then mixed with water that instantly draws the ionic polymer species that resulted into the aqueous medium. On removal of the organic solvent, a stable dispersion of 40%–50% solids having different trends and degrees of crystallinity, free of the surfactant, may be obtained. Dispersions of PU so formed may be blended with other polymeric dispersions, giving hybrid adhesives of good to excellent bond strength. Environmental concerns attach great importance to the development of such *solvent-free adhesive formulations* in aqueous dispersions.

4.8 Acrylic Adhesives and Cyanoacrylates

Acrylic adhesives are commonly based on polymers or copolymers of selected acrylate and methacrylate monomers. They are available both as solvent-based or aqueous emulsion-based adhesives. Acrylates having more than four carbon atoms in the ester-alkyl group yield polymers with low T_g; methacrylate polymers have a higher T_g than the corresponding acrylate polymers. Acrylate copolymers of T_g ranging between –60°C and 30°C and exhibiting high tackiness are popular as adhesives; using acrylic acid as a comonomer improves adhesion, in general. Such acrylate adhesives are photocurable; cross-linking can be effected using di- or poly-isocyanates, phenolics or epoxy-phenolics and amino resins. Polymers/copolymers based on N-methylol acrylamide or butylated N-methylol acrylamide are heat cross linkable. In acrylate copolymer synthesis, using 2%–5% of N-methylol acrylamide or acrylic acid is helpful, because they act as the cure-site monomeric units and render the copolymer adhesive selectively curable.

Acrylates are widely used in pressure-sensitive adhesive formulations. They stand second and next to the top rated elastomer-based materials in this application. Common monomers for making acrylate adhesives are: 2-ethyl hexyl, butyl and octyl acrylates; low doses of vinyl acetate and methyl methacrylate and minor proportions of acrylic acid and acrylamide are suitably used as comonomers. Even without cross-linking after applications, high molecular weight acrylate polymers/copolymers used as pressure-sensitive adhesives exhibit adequate cohesive strength.

Cyanoacrylates are used in monomeric form and they polymerize in the bond line. Methyl and ethyl cyanoacrylates ($CH_2{=}CH \cdot COOCH_2CN$, $CH_2{=}CH \cdot COOCH_2 \cdot CH_2CN$) are used. Because of the presence of two closeby electron withdrawing groups (C=O, and C≡CN) on the substituent ester moiety, the monomers undergo anionic polymerization even under the influence of ambient moisture (OH^- anion) in seconds or minutes. Curing is accelerated by amines or ammonia on acidic surfaces. (Sulfonic) acids impart some stability to the liquid cyanoacrylate monomers.

Cyanoacrylate glue-lines show poor heat resistance, hydrolytic instability and prominent brittleness. Copolymerization with methyl acrylate provides toughness. If duly formulated, the cyanoacrylates find wide acceptance in the consumer market due to their rapid curing and good adhesion to ceramics, glass, metals, plastics and elastomers. They are not highly rated in large scale industrial applications.

4.9 Silicone Adhesives

The use of silicones as sealants exceeds their use as adhesives. They are favoured because of their excellent physical properties, high elongation, exceptional chemical inertness, stable viscosity over a long temperature range and their heat resistance. In pressure-sensitive adhesives, these properties, along with excellent adhesion to many surfaces, hold high prospects for their growth in the future.

Sealants are produced from polysiloxanes that cross-link in the presence of moisture with a tri/tetra-functional silane possessing readily hydrolyzable groups. Silicone pressure-sensitive adhesive manufacture involves the reaction of a gum (elastomer) and resin. The gum is based on dimethyl or blends of dimethyl and diphenyl or methyl-phenyl linear polysiloxanes. Molecular weights of 80,000–1,20,000 are useful. Resins are made by the reaction of trimethyl chlorosilane or hexamethyl disiloxane with silicic or polysilicic acid in the presence of a weak acid and a solvent. The resin is essentially a condensation product of alcoholic silica hydrosol and monofunctional (chloro) silane in a predetermined proportion of reactive functional groups, giving $\bar{M}_n$ of ≤ 1000. For making adhesives, the resin is dissolved in a solvent and the gum is then added in pieces with agitation. The resin acts as tackifier for the gum component. The resin to gum weight ratio is maintained at 60 : 40.

4.10 High-temperature Adhesives

A selected group of polyheterocyclic polymers, e.g. polyimides, polybenzimidazoles and polyquinoxalines are potentially attractive as adhesives for long term, high temperature aerospace applications. They are favoured for their retention of high bond strength for thousands of hours at temperatures ≥ 300°C. The substrates bonded are commonly titanium and carbon fibre laminates. The adhesives are also distinctive for their excellent resistance to oxidative degradation and formation of high char residues. Their high price and big processing difficulties come in the way of their widespread adoption, uses and applications.

4.11 Inorganic Adhesives

Soluble silicates made by the fusion of purified silica with alkali carbonate are acknowledged as the most important inorganic adhesives. The most dominant form of soluble silicate is sodium silicate with a SiO_2–Na_2O ratio of 2 : 1 to 3.7 : 1. The silicates commonly set to a hard mass simply by drying. Clay is often mixed to control shrinkage. Other additives of importance are:

(a) starch to impart a thickening effect
(b) protein glues to provide water resistance
(c) surfactants as wetting agents, and
(d) humectants to delay setting and enhance tackiness

Silicate adhesives are widely used in paper bonding, corrugated board and box industries.

4.12 Fusion Bonding

This is a process that involves simultaneously heating the bond surfaces on the edges of two substrates, bringing them close together while the edges are still in the molten state and applying sufficient pressure on contact face to face to compensate for thermal contraction and allowing the mutually-fused bond to cool. It is the most common method of joining plastic (PE) pipelines for natural gas. To be fusion bonded, both the substrates must be thermoplastic, melt compatible, (nearly) identical in thermal expansion, preferably similar in thickness and diameter, and equal in shrinkage on cooling.

Heat sealing thermoplastic pouches is viewed as indirect fusion bonding, because the heat is applied at the outer surface and conducted into the bond interface. Heat sealing of polymers having sharp crystalline melting points such as nylon and PET are much more difficult than heat sealing of PE.

On cost/equal area considerations, fusion bonding and mechanical bonding (e.g. insert moulding, potting, hot dipping and clinching) are more productive than fastener or adhesive bonding. "Insert moulding" is simply placing a substrate (usually metallic) in a mould and covering part or all of it with a second material, commonly a liquid oligomer or a polymer (plastic) melt—plastic handles are insert-moulded on screw driver blades. Clinching is a permanent mechanical bonding technique requiring one of the substrates to be easily pierced and stretched or to be malleable enough to deform when stamped. Mechanical bonding does not require that separate inventories of expensive fasteners or adhesives be maintained.

4.13 Pressure-sensitive Adhesives (PSA)

Some adhesives have the special property called *adhesive tack,* i.e. the property of sticking quickly and firmly to a surface after a brief contact under a low, positive pressure. They wet the surface quickly like a mobile fluid and yet, resist easy detachment, as if they have been transformed rapidly into a cohesive solid. These features, together called *autohesion,* are apparently contradictory and are achieved by carefully controling the physical properties of the adhesive. First, to secure rapid wetting, the adhesive must show a small-time (immediately on contact) *compressive creep compliance* greater than 10^{-6} m^2/N. For compliance greater than this value, the forces of attraction between molecules of the adhesive and the adherend substrate are high enough to pull the adhesive layer into intimate contact with the adherend substrate surface, despite the surface being irregular on a microscopic scale.

A substance that readily flows at low stresses but hardens at high strain is potentially ideal as a pressure-sensitive adhesive. Some elastomeric systems are self-strengthening on deforming, by virtue of the steric regularity of their chain molecules (1, 4 cis-polyisoprene or natural rubber, poly (trans chloroprene), polyisobutylene) and they will all crystallize and stiffen at high strains; they have all been employed in pressure-sensitive adhesive formulations. Such formulations possess little resistance to small strain deformations, thus enabling rapid wetting, but they are required to support large strains without flowing. Loose networks of high molecular weight polymers of high chain entanglements are most suited for this, particularly if duly mingled or diluted with a selected dose of specific types of *tackifiers,* such as ester gum (rosin ester of glycerol), coumarone-indene resin, terpene oligomers and resins, aliphatic petroleum resin and alkyd-modified phenolics.

4.13.1 Tackifiers

Using tackifier resins (molecular weight 500–2000, and $T_g \geq 40°C$) favourably modify the rheological attributes of an elastometric formulation for adhesive application. Resins commonly used as tackifying agents are: rosin and rosin derivatives (limed rosin and ester gum), dimer acid-based polyamides, coumarone-indene resins, terpene oligomers/resins, petroleum resins, and alkyl-modified phenolics. When the resin is used, the resistance to deformation at lower rates gets lowered; hence facilitating bond formation on contact. Mechanical testing further indicates a higher storage modulus (G') at higher rates of deformation, as in Fig. 4.3, and the bonded material appears stronger.

Using a filler material raises the G' value over the full range of deformation rate, thus hindering bond formation, while using a simple plasticizer expectedly lowers the cohesive strength of the adhesive. It appears then that a tackifier is a genuine compromising agent to enhance bond formation and improve elastic modulus; as a consequence, it imparts a prominent reinforcing effect at the time of detachment.

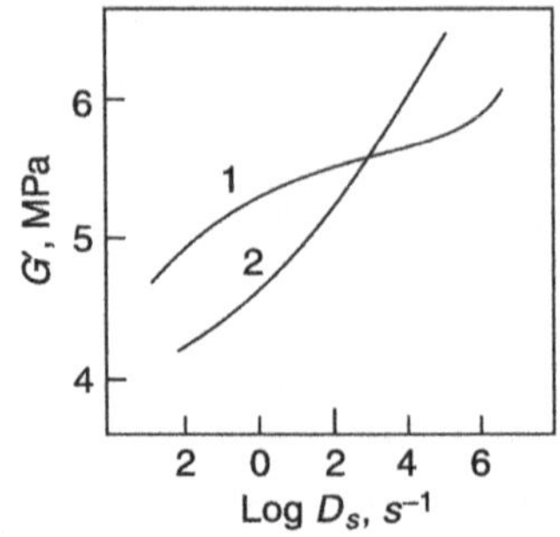

Figure 4.3 The effect of the use of a tackifying resin on storage modulus (G') of natural rubber: (1) using no resin; (2) using ($\leq$ 10 phr of) a tackifying resin

Pressure-sensitive adhesives (PSAs) are commonly soft elastomeric semi-solids. Their peel strength (τ) is strongly contingent upon the rate of peel (R_p) and the test temperature, as in Fig. 4.4. At a given temperature, the bond failure mode shifts from *cohesive fracture mode* at a low rate of peel to *interfacial fracture mode* at a relatively high rate of peel. Below the critical rate, the peel strength is more a measure of the work of extension of the viscous/viscoelastic rubbery mass to the point of rupture; above the critical rate, the peel strength is only dependent upon interfacial interactions and dissipation forces within the adherends. For two elastomeric layers to adhere well together after contact, the surfaces in contact must wet each other through the in-built property of tack and must resist high stresses without exhibiting flow.

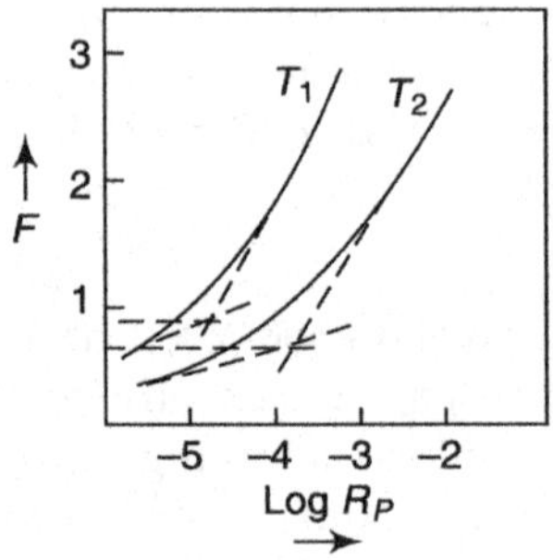

FIGURE 4.4 Trend of the variation of peel force *(F)*, kN/m vs log (rate of peel, R_p), m/s at T_1 = 10°C and T_2 = 30°C for NR-based pressure-sensitive adhesive

4.13.2 Autohesion and Green Strength

Amorphous elastomers of high molecular weight having little tendency to crystallize on stretching rely on chain molecular entanglements for providing *green strength* and *cohesion*; they also restrict the chances of wetting on contact for restrictive flow characteristics. For a relatively low molecular weight elastomeric system, there is low scope for chain entanglement; this results in some degree of flow with relative ease and hence low green strength. Natural rubber (NR) being strain-crystallizable shows much improved autohesion. Mastication of NR lowers viscosity, enhances flow, confers wetting and still shows high-to-moderate green strength, depending on the degree of mastication, owing to *strain-hardening* consequent to strain-induced crystallization. For an elastomer, to be partially crystalline in the unstrained state is less than desirable in PSA applications, as this would seriously hinder wetting and bond-formation on contact.

The *Autohesion* of an elastomer is interchangeably referred to as *tack* and the upper limit of autohesion or tack is taken as the *green strength* of the elastomeric adhesive compound. NR, SBR, PBR, butyl rubber and their selected blends are used in pressure-sensitive adhesive formulations. The ratio of tack to green strength, usually a fraction, is commonly referred to as *relative tack* and its maximum possible value is unity (1.0). Over the molecular weight range of 1×10^5–5×10^5, the (absolute) tack nearly equals green strength and the relative tack is close to unity. With further increase in molecular weight, molecular mobility, wetting and diffusion are progressively restricted and relative tack value gets progressively lowered, deviating more and more from unity.

Pressure-sensitive adhesives find wide application as adhesive tapes, viz. cellophane (Scotch) tapes, surgical tapes, insulation tapes for electrical application, masking tapes, etc. Their function is to provide a temporary bonding action with the ability to be easily removed after completing the function they are intended to discharge without damage to the substrate. It is not desired or necessary for these adhesives to ever set or cure to a hard glue-line.

4.14 Sealants and Mastics

A group of materials called *sealants* or *mastics* are often classified with adhesives, even though they have different functions. Adhesives are used in low thicknesses for their high ability to bind two surfaces. Sealants are selected as load-bearing, elastic-jointing materials primarily to exclude entry or inflow of dust, dirt, moisture or chemicals (that may contaminate a liquid or gas). Sealants are used as relatively thick layers that not only seal, but also reduce noise propagation and vibration and act as insulators or space fillers.

The term *caulking compound* relates to the verb "caulk" and initially, a caulking compound meant tarred oakum (strongly adhering putty-like substance) forced between the planks of a boat or ship. Initially they were mixtures of bitumen and asbestos meant to fill and seal and had little to do with load-bearing. The sealant is a modern term for caulking compounds and modern sealants are usually compositions based on synthetic resins, asphaltic and oil-based, water-repellent, non-permeable materials. Modern sealants are mostly based on polysulfide rubbers, silicones (liquids, greases, resins and rubbers), butyl rubber, polyurethanes, acrylics, polychloroprene, chlorosulfonated polyethylene, bitumen, PVC-plastisols, polyesters, phenolic/urea resins, epoxy resins, furan resins, etc.—all duly compounded with fillers, plasticizers, flow-promoters, antioxidants, colourants, curing agents (if necessary), initiators or catalysts, accelerators and activators. Compositions based on butyl rubber, bitumen, EVA, atactic polypropylene, polyolefin blends and blends of paraffins and polyolefins are used as hot melt sealants. Mastics based on oil-modified alkyd resins and alumina find special use as linings and gaskets, and in laying or repairing refractory bricks, plates or linings in metallurgical process units and furnaces. Bitumen-asphalt mastics are used in laying and repairing road surfaces.

A two-pack sealant based on silica or carbon black filler containing p-toluene sulfonic acid (10%) in one pack and high-methylol phenolic resole (A-stage) resin in the other pack makes a good room-temperature cure sealant. The acid acts as the curing catalyst when the two packs are mixed together intimately and applied as a sealant, while still in liquid or dough form, to seal linings between bricks or tiles. A typical formulation for a corrosion-resistant epoxy resin sealant consists of 100 parts liquid epoxy prepolymer, 50 parts liquid poly (ethylene sulfide) and 10 parts dimethyl amino propyl amine (curative), optimally admixed with selected fillers (10–30 parts).

References

1. Schmidt, A X and C A Marlies, *Principles of High Polymer Theory and Practice*, McGraw-Hill, New York, 1948.
2. Mark, H F, N M Bikales, C G Overberger, G Menges and J I Kroschwitz, eds, *Encyclopedia of Polymer Scinece and Engineering*, Vol. 1, John Wiley and Sons, New York, 1985, pp 476–546.
3. Roberts, A D, *Adhesion—1*, K W Allen ed, Applied Science, 1977, p 207.
4. Krupp, J and W Schnabel, *J Adhesives*, Vol. 5, 1973, 296.
5. Fowkes, F W, *Treatise on Adhesion and Adhesives*, New York, R L Patric ed, Marcel Dekker, 1967.
6. Rossman, K, *J Polym. Sci.*, Vol. 19, 1956, 141.
7. Kim, C Y and D A I Goring, *J. Appl. Polym. Sci.*, Vol. 15, 1971, 1357.
8. Ghosh, P, *Polymer Science and Technology—Plastics, Rubbers, Blends and Composites*, 2/E, Tata McGraw-Hill, New Delhi, 2002, Ch. 9.

Chapter

5

Coatings: Principles and Practice

A continuous cohesive cover in the form of a film of different thicknesses (nanometer or micrometer range to ≤ 1 mm) spread on the surfaces of flexible substrates (paper, fibre, textile, leather, metal foil, wire, plastic films, etc.) or rigid substrates (sheets, strips, tubes, pipes, blocks of wood, bamboo, glass, metals, ceramics and plastics), providing protection, comfort, decoration and durability may be commonly called a *coating*.

Coatings are also being provided to fine drops of specified liquids and emulsions and to powdery or granular particles of specified solid chemicals, drugs and pharmaceuticals, fertilizers, pesticides and the like, to impart pressure-release (copying paper) or control-release characters to meet technology needs and for efficiency in material use, to minimize wastage and loss of potent materials and for working efficiency along with economic advantage.

5.1 Fundamental Types of Surface Coatings

Saving a surface is as important as, or even more important than, making the surface. The two main functions of surface coatings[1–3] are decoration and protection, and in most surface coatings these functions are combined. Organic resins and polymers, mostly of synthetic origin, constitute the base (film forming) materials for most surface coating formulations. They are commonly applied as fluids (solutions, dispersions and emulsions) and must dry to thin, continuous, adherent, cohesive, solid films. Contingent upon the mechanism involved in the drying process, they may be classified into two fundamental types: (a) lacquer-type and (b) varnish-type.

5.1.1 Lacquer-type Surface Coatings

Lacquer is the name for coatings distinguished by a drying characteristic involving merely the evaporation of the solvents and thinners present in the coating formulations. The drying does not involve chemical reactions. All polymeric constituents of a lacquer formulation are present in fully polymerized form at the point of application. Lacquer-drying is not a reactive way of drying.

5.1.2 Varnish-type Surface Coatings

In a *varnish* formulation, drying after application involves not only evaporation of the solvents and thinners, but polymerization, chain extension, oxidation and curing reactions as well, which ultimately lead to the formation of molecules in the dried film that are larger, more complex and variously different from those present at the point of application. Drying of a varnish formulation is, in fact, a case of reactive drying along with (drying by) solvent evaporation.

Lacquer-type surface coating formulations are usually quicker drying than the varnish-type surface coating formulations. Surface coatings may be further sub-classified into (i) *air–drying types*, allowing drying under room–environment conditions and (ii) *baking types*, that require elevated temperatures for setting to finished, dry and durable films.

5.1.3 Paints and Enamels

We may also have a sub-classification of surface coatings as (i) paints and (ii) enamels. A paint is a mixture of semi-opaque or opaque adhering substances with liquids (solvents/dispersants) that render them suitable for spreading over surfaces by different application techniques like brushing, dipping and spraying; they are spread with ease and eventually, they dry to a durable, thin coating. An enamel, however, typically dries to a durable film of excellent to outstanding hardness. Paints and enamels may be either the lacquer- or the varnish-type. A typical paint is generally low in or devoid of hard resin constituents, while a typical enamel essentially contains appreciable amounts of hard resin constituents; there is, however, no sharp dividing line between paints and enamels.

5.2 Lacquers

A lacquer-type surface coating arises from a simple formulation consisting only of one or a blend of film-forming resin(s) or polymer(s) and a volatile solvent or solvent mixture, a simple example being French polish or furniture polish, or (wooden) floor polish, containing shellac dissolved in denatured alcohol or methylated spirit. A great majority of lacquer-type coatings are based on cellulose nitrate and ethyl cellulose (10–15 parts) and minor to moderate (5–30 parts) extents of some other resins (ester gums and plain or rosin/maleic modified alkyd polyester resins) acting as binders additionally. They may contain a permanent external plasticizer (dibutyl phthalate 3–5 parts), selected solvents and diluents (60–80 parts) and optionally, some colour and pigment (1–2 parts). Since no chemical reaction is involved while drying, there is no need for a catalyst.

A spurt in the automobile industry in the initial decades of the twentieth century led to development of cellulose nitrate-based lacquers. To be effective and sprayable, the viscosity of the solution needed to be kept below 70 centipoises while maintaining the concentration of cellulose nitrate (coating grade, having 11%–12% N_2 content and D P, 170–180) at 10%–15%. Coating-grade cellulose nitrate of this D.P. and concentration provides a desirable solution viscosity for application, giving optimum solids content and surface coverage when blended with appropriate lacquer resins, which meaningfully contribute to improve adhesion, gloss, hardness and durability. Along with cellulose nitrate, other cellulose derivatives such as ethyl cellulose and cellulose acetate-butyrate are popular as lacquer binders. The combination of appropriate cellulosics as binders and selected plasticizing resins offers a better property range and property balance than either one alone. The resins used have T_g of 15–40°C; they increase the total solids content of the lacquer mix without notably increasing its viscosity, since the resin molecules are not very large. In their presence, coatings of the desired thickness and coverage are more easily attained. They confer and enhance adhesion and resistance to light, moisture and weather. They adequately plasticize cellulose nitrate and relevant cellulose ester/ether binders—in this respect, they appear more effective than liquid external plasticizers. They also resist possible loss by migration to the surface and slow evaporation.

Among the lacquer-plasticizers, those known as the solvent type and having complete miscibility with cellulosic and resin binders include dibutyl phthalate, tricresyl phosphate, triphenyl phosphate and camphor; they produce uniform dispersions or solid solutions. The *non-solvent* plasticizers with limited miscibility with the binder include blown castor or soybean oil, boiled linseed oil and tung oil. Results are improved by using a 2 : 1 (w/w) blend of solvent-type and non-solvent type plasticizers.

5.2.1 Controlling Features in Lacquer Formulation

One of the most important balancing considerations in the lacquer field is to design a formulation having a viscosity low enough for trouble-free brushing or spraying, yet giving a solids content high enough to produce an effectively thick film with a single or few coats, thus offering process as well as cost efficiency. The steps to be ensured include using:

(a) low viscosity or low/moderate chain length (DP of 170–200) binders (cellulose nitrate cellulose acetate butyrate, ethyl cellulose, etc.)
(b) large percentages of plasticizing, low (2000–5000) molecular weight resins
(c) the method of hot spraying
(d) improved spraying equipment capable of manipulating relatively viscous materials, and
(e) solvent systems or vehicles of high solvent power

Refer to the relevant sections in Chapters 1 and 2 concerning solvents and diluents for polymers, their swelling, solubility, flow character, viscosity and sol-gel transformations covering thixotropy and dilatancy.

Solvents variously included in lacquer-formulations generally fall into three major categories:

(i) *Active solvents*: Solvents that normally produce solutions of low viscosity at high binder concentrations and dry up to give cohesive films, free from shortcomings such as pinholes, wrinkles, waviness or 'orange peel'. Esters (ethyl acetate, amyl acetate), ketones (acetone, isobutyl ketone) and alcohol ethers (ethoxy/methoxy ethanol, i.e. ethyl cellosolve or methyl cellosolve) act as active solvents

(ii) *Latent solvents:* Solvents that are themselves unsatisfactory lacquer-solvents, but when present, enhance the solvent power of the active solvents by improving solubility and lowering viscosity. Alcohols (ethanol, isopropanol, butanol and amyl alcohol) in effect act as latent solvents

(iii) *Diluents*, such as aliphatic or aromatic hydrocarbons (hydrogenated petroleum naphtha, toluene and xylene), which by themselves are not lacquer solvents but when used in the right proportions with active and latent solvents, effectively contribute to lower the viscosity, advantageously modify evaporation rates and meaningfully reduce solvent costs.

5.3 Adhesion and Application Concerns

Adhesion of a surface coating is virtually dependent upon adequate *preparation of the surface* to which it is applied, while its performance and durability are influenced by the *manner* in which it is applied (see also Chapter 1). For every surface coating, there is a film thickness that develops optimum mechanical properties and durability. Inappropriate thickness, higher or lower than the optimum, lowers the service life. A coating from the same formulation at two different film thicknesses may exhibit altogether different types of failure. This factor is of particular importance in varnish-type coatings (see Chapter 3). If the coating is too thin, weakness and rapid breakdown would result whereas too thick a film would result in poorer and delayed solvent release, causing wrinkling, waviness and other surface defects.

Wrinkles normally develop in a region of stress. Film wrinkles would finally lead to the development of cracks. Cobalt driers tend to support the formation of deep wrinkles with strong cracking tendencies. Manganese and lead driers however, produce improved films in this respect.

5.4 Pigments and Pigment Dispersion

A pigment is a finely ground, insoluble particulate matter that imparts colour (including white and black) and opacity to paints, enamels, varnishes and lacquers. The role of a pigment is that of a filler and colouring matter for the final surface coating film. Some of the important pigments are non-reactive, while some such as oxides of lead and zinc and whiting are alkaline. As *vehicles* (liquid composed of binder and thinner) of many surface coatings contain acidic components, chemical reactions between them and the alkaline pigments invariably take place with the resultant formation of heavy metal soaps and related compounds; if prominent, it may ultimately cause the film to deteriorate. As varnish-type coatings generally contain *reactive binders* as opposed to the lacquer-type coatings, it is logical to conclude that storage-time or post application chemical reactions play a significant role in the former. Exposure tests have established that the life of exterior paints is longer with duly selected combinations of pigments and extenders than when one type is used alone. In black formulations, exterior paints or inks, selected grades of carbon black are the invariable choice of pigment.

Wetting of the ground pigment particles is absolutely essential for obtaining good grade coatings. Poor or limited pigment wetting may cause poor pigment dispersion and rapid settling, and rapid caking may result in the container.

5.5 Paint Manufacturing

The manufacture of paint involves mixing, grinding and thinning using solvents, plasticizers and thinners or diluents, filling the paint into a container, labelling and shipping it. Grinding is the most critical step here. The term "grinding" is a misnomer, as no size reduction or breakdown of pigment particles occurs during this process in paint manufacture. In fact, the actual grinding process is involved during pigment manufacture itself, and the pigment grains/agglomerates are reduced to particles of the proper size and fineness. However, "grinding" in the context of paint-making consists of wetting and dispersing the pigment particle agglomerates into single particles or finer agglomerates in the paint vehicle by working them in a high-speed, rotating (pebble or steel ball) mill, effectively removing the harmful air and moisture envelopes on the pigment particles. High-speed grinding in the dissolver/disperser completes the *pigment dispersion* efficiently and quickly, at a low cost.

Sometimes, the fine pigment particles that are basic in character are treated with a dilute solution of a long chain fatty acid (lauric acid, palmitic acid, stearic acid); subsequently, removing the volatile solvent results in the formation of *activated pigments*. Now, a very stable film of oriented fatty acid molecules surrounds the pigment particles with the carboxyl groups directed toward the pigment surface and the paraffinic hydrocarbon chain portions are pointed or oriented outward. This oriented layer not only acts as a lubricant hindering the formation of agglomerates of pigment particles, it also renders them prone to ready wetting by the vehicle containing the binder dissolved in the solvents and diluents. This pre-treatment of the pigment also hinders possible adsorption of drier (Co, Mn, Cu, Pb) soaps by the pigment, thereby preventing possible impairment of the desired drying capacity of the paint and related coating formulations on application. By varying the pre-treatment, the same pigment may be rendered dispersible in organic solvents or aqueous media. Water-based paints need to be formulated using water-dispersible pigments. The advantages of water-based paints are:

(1) Elimination of odour, fire hazard, and toxicity; lowering of solvent cost
(2) Applicability on damp surfaces
(3) Greater ease and lower cost of application
(4) Greater drying speed (second coats are applicable in 1–3 h)
(5) Production of a uniform, flat finish and
(6) Outstanding bright colours with little tendency to yellowing.

Many pigments are either precipitated or washed in water during their preparation. The growth and popularity of water-based paints, offering many advantages, has led to the direct use of water slurries of pigments in paint manufacture. Using slurries allows bulk storage, easy handling and advantageous shipping using pumps and pipe lines, and simplifies the process of making such paints. Modern paint plants usually utilize slurries of titanium dioxide and common extender pigments. Titanium dioxide, a white synthetic pigment is the most common; it has a relatively high refractive index (2.5–2.7) and is hence efficient in scattering light or providing good *hiding power* (See Table 5.1). Coloured pigments having higher hiding powers are usually metallic oxides (oxides of iron, chromium, lead, etc.) or synthetic organic pigments like phthalocyanin blue, Hansa yellow, etc. Colour (pigment) concentrates or master batches are prepared for uniformly mixing with the white base paint to make paints of a specific colour and shade.

TABLE 5.1 Approximate refractive index values of some white pigments

Pigments	*Refractive Index*
Titanium dioxide	2.5–2.7
Zinc oxide	2.01–2.1
Zirconium oxide	2.4–2.5
Zinc sulfide	2.4–2.5
Antimony oxide	2.2–2.3
Basic lead carbonate	2.0–2.1
Basic lead sulfate	1.8–1.9
Barytes	1.6–1.7
Calcium carbonate (Chalk)	1.5–1.6
Calcium sulphate	1.5–1.6
China clay	1.5–1.6
Silica (Quartz)	1.5–1.6
Talc	1.5–1.6

5.5.1 Gloss

The thinner, or volatile matter, of latex and emulsion paints is invariably water. In non-latex paints, the volatile solvent is usually mineral spirit. In most paints, the property that varies most, other than colour, is *gloss*. Gloss is the reflection of light from a surface to the same angle as the angle of incidence. Thus, planar surfaces are necessary for glossiness. A paint binder with no pigment normally dries to a smooth, glossy finish. There is little disturbance of gloss for low pigment concentration, as the pigment particles remain embedded in the binder layer. With the use of more pigment, the pigment particles tend to protrude from the binder surface, diffusing the incident light and impeding true reflection; this results in low gloss.

When the binder is just sufficient or insufficient to wet the pigment and fill the voids, gloss suffers severely and falls to the bare minimum. The percentage volume of the dry film occupied by the pigment is called *pigment volume concentration (PVC)* and the transition area corresponding to minimum gloss is the *critical pigment volume concentration (CPVC)*. Poorly balanced solvent mixtures in lacquers may cause surface irregularities. A vortex air current set up during solvent evaporation may cause surface defects called *orange peel*. Moisture absorption by lacquer films may cause the development of turbidity, called *blushing*. Blushing can be avoided by keeping the free glycerol content and acid value low; the acid value should preferably be < 20.

5.5.2 Primers

Paints for the first coat or *undercoat* for metals, walls and wood are called *primers*. Top coats are *finishing coats*, giving flat, semi-glossy to glossy upper coats. Primers are also called *sealers* as they seal the pores on the surface of the substrate and adhere to it and to subsequent paint upper coats. For most architectural uses, alkyd-based primers or sealers are preferred.

5.5.3 Pigment Loading and Flow Properties

The flow behaviour pattern of paint is an important quality parameter. Paint must be thixotropic; this point has been dealt with in detail in Chapters 1 and 2. The thixotropic set to a gel structure imparts sufficient body during storage to minimize pigment settling and avoid caking. However, the paint should attain desired flow upon shearing appropriately and flow under brush; it should set again as brushing ceases at a rate that gives the brushed film enough time to level out the brush marks. Higher concentration of suspended pigment particles increases the consistency and yield point, as in Fig. 5.1. The oil of varnish-type paints is apparently Newtonian in flow character, but as the dispersed pigment content is increased, the paint exhibits increasing yield value and lower flow rate.

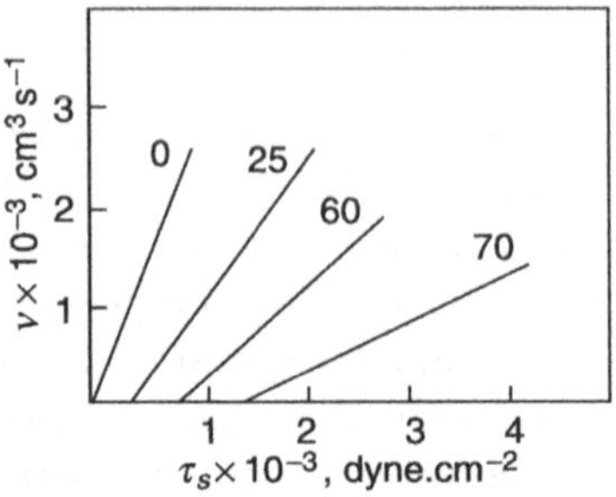

FIGURE 5.1 Plot of velocity ν vs shearing stress τ_s showing the effect of increase in pigment loading (from 0%–70%) on flow behaviour and yield value (given by the intercept on the abscissa) of linseed oil-based, varnish-type paint

5.6 ELECTROPLATING

Electroplating[3,4] relates to the electrodeposition of an adherent metallic coating on an electrode to form a surface with properties different from those of the base metal. The term "electrodeposition" is used to also cover electroless plating and immersion processes that do not employ an electric current.

The technological and practical phases of electroplating came to maturity much before the understanding of the relevant science and controlling principles involved. Electroplating copper, silver and gold dates back to the days of Faraday when he enunciated the laws of electrolysis. In the early phase, electroplating was more an art than a science and the electroplaters had little knowledge of electricity, ion dissociation and ion discharge. Electroplating initially started as a decorative coating process and the appearance of the plated product was more important than its physicochemical properties.

Electroplating, in effect, is a surface treatment process. The material (usually a metal) being treated is made and used as the cathode (negative electrode) in an electroplating solution bath. Such baths almost always contain aqueous solutions such that only those metals appropriately placed in the electrochemical series that could be reduced from aqueous solutions of their salts can be electroplated. An exception is aluminium (Al), which is plated, in a semi-commercial scale from organic electrolytes.

The thickness of the metal deposit applied by electroplating is guided by application standards, viz. 0.025 μm for gold deposits, 20–50 μm for standard nickel-chromium plating for, say, exterior automotive hardware and ≥ 1 mm for various electroforms.

The advantages of electroplating are:

(i) Improved corrosion resistance
(ii) Attractive appearance
(iii) Improved frictional characteristics
(iv) Higher wear resistance and hardness
(v) Solderability,and
(vi) Some desirable and specified electrical properties etc.

Electroforming is used to manufacture articles that can not be made as economically in any other way. Consumption of different metals by electroplating is in the order: tin ≥ nickel > chromium >>> silver >> gold.

The final articles from electroplating consist of the surface deposit or plating and the base metal or substrate. It is the properties of this combination that ultimately determine the right metal to plate and the right solution to plate it from.

Many metals, corrosion resistant themselves, actually decrease the corrosion resistance of a composite when they are plated over a less noble metal. The galvanic couple set up by the contact of two dissimilar metals promotes corrosion. This point is of particular importance in plating upon aluminium, which is electrochemically more active than many metals with which it can be plated. Thus, the procedures for plating on aluminium must be chosen and carried out with the utmost care. The plating process may impart an adverse effect on the substrate—such as the embrittlement effect of hydrogen liberated on high strength steel, particularly in zinc and cadmium plating.

Further, the possibility that the plating may diffuse into the substrate with time, forming an alloy that is apparently favoured at elevated temperatures, should be borne in mind. Such alloy formation may develop brittleness and other undesirable properties. To avoid such negative effects, it is better to interpose a barrier layer of another metal, most often nickel, between the substrate and the final plate. For tin-plating on brass for soderability, standard specifications prescribe laying a barrier coating of at least 2.5 μm of copper, bronze or nickel to prevent the migration of zinc to the surface.

5.6.1 Surface Conditioning for Plating

Before a useful electroplate can be deposited on a surface, the surface must be in the right condition to receive it. The pre-plating surface treatments commonly include cleaning, i.e. pickling to remove gross scale, polishing or buffing to generate a fresh surface. This sequence is called "mechanical preparation;" further cleaning follows to remove oil, grease and dirt applying polishing/buffing compounds and rinsing. Additionally, the surface is dipped in acid to remove any oxide film on it and rinsed. Organic solvent cleaning helps remove oils and greases, and a mild alkali cleaning follows the organic (solvent) cleaning. Adequate rinsing between all steps in cleaning and plating is of great importance. If plating is to be done in an alkaline solution, it is desirable to use an alkaline dip following the rinse after the acid-dip. In general, when work is to proceed from an acid to an alkaline solution or vice versa, it is worth inserting a neutralizing dip in between.

5.6.2 Plating on Nonconductors

Electroplating on nonconductors like wood, leather, plastics, etc. requires a conducting surface with at least minimal adhesion to the substrate. The processes employed to achieve this include graphitizing the surface, applying bronze or other metallic paints, or chemical silvering by mixing a silver salt solution and a reducing solution from a two-nozzle gun so that the solutions mix just as they hit the surface to be silvered. Such methods, however, result in poor adhesion.

Today, plating on synthetic plastics is a widely adopted and developed technology. The cycle for preparing the surfaces of these plastics (ABS, PP, polysulphones, polycarbonates, etc.) includes: (cloth) cleaning to remove soil, dust and dirt, oil/grease and fingerprints; solvent treatment and conditioning to render the surface catalytic and applying an electroless or autocatalytic deposit, usually nickel or copper. Conditioning, commonly done with a solution of chromic acid and sulphuric acid, roughens the surface just enough to yield interlocking of the subsequent deposit to be formed and changes the surface chemically to provide sites for chemical bonding. The catalytic surface is generally provided by the adsorption of a reducing agent, normally stannous chloride ($SnCl_2$), followed by immersion in a dilute solution of palladium chloride, which is reduced to metallic palladium by the Sn^{2+} ion on the surface. A one-step process containing both stannous chloride and palladium chloride in one solution can be used to provide a conducting surface, after which electroplating takes place as usual. Plating on plastics is a vital step in the manufacture of printed circuit boards; this has grown and is still growing in use and importance. Waste disposal and metal recovery are of prime concern and importance in the context of pollution control imperatives and environment preservation.

5.6.3 Alloy Plating

This relates to the simultaneous plating of two or more metals. It is viewed as an attractive field of research and development even though few processes have been commercialized. There is a lot of potential for growth in this technology

The main problem with alloy plating is the need for additional control. Changes in operating variables such as temperature, current density and bath composition would affect the relative rates of deposition of two or more depositing metals, thus necessitating the maintenance of process variables within narrow ranges, which is not required in single metal plating. Complications in practical alloy plating increase almost exponentially as the number of codepositing metals is increased. In selective cases, alloys can be formed by allowing the deposition of more than one metal separately and subsequently interdiffusing them by applying of heat, even though such processes are far from true alloy plating.

Brass is recognized as the first commercially plated alloy; its most important engineering use is to ensure good adhesion of rubber to steel bead wire in tyre manufacture. Most commercial baths are of the cyanide type with wide variations in composition. True bronze is a copper-tin alloy, even though in the electroplating industry, "bronze" may mean alloys of copper with zinc, cadmium or tin, matching the colour of wrought bronze.

There are many other ways than electroplating to deposit a coating of a metal on some other specified metal or other substrate. Hot dipping, vacuum evaporation, chemical vapour deposition and various other processes not requiring electric current have gone through rapid developments in recent years. Non-electrolytic aqueous deposition includes (a) immersion plating (see section 5.7) and (b) chemical, autocatalytic or the more familiar electroless coating (see Section 5.8).

5.6.4 Electro-deposition (ED) or Electro-coating (EC)

Electro deposition, or electro coating, is a process of depositing an organic polymer-based coating or paint on an electro conductive (metallic) substrate from the aqueous solution of the polymer-based paint formulation under the influence of an electric DC field. Depending on the nature of the charge carried by the polymeric binder, the coating occurs either on the cathode or on the anode. Anodic electro deposition (AED) refers to the migration of negatively-charged (usually carboxyl, COOH-bearing) paint particles on the positively-charged (electrode) substrate used as the anode, whereas cathodic electro-deposition (CED) refers to deposition of positively-charged (quaternary ammonium/amine/amide-group bearing) paint particles on the negatively-charged substrate used as the cathode.

Electro deposition or electro coating is in wide practice for imparting the primer coating on charged car body parts and is accomplished by immersing the parts to be coated in an electro deposition tank containing a stable aqueous suspension of the primer paint; a DC voltage is applied from a rectifier for a specified time period. Depending on the charge carried by the film-forming polymer, the paint is cathodically or anodically deposited as a compact, semi-dried film on the car body.

A thin loose, non-adherent layer of paint covering the electro deposited base film is also formed and is subsequently removed by rinsing. The semi-dried, wet, electro deposited film is further dried and hardened by cross-linking at an appropriately high temperature. The electro coating process has been developed as an automatic mass production technology; it involves minimum lossses and is less labour-intensive, thereby providing economic benefits to automobile manufacturers.

Advanced cathodic binders, mostly derived from diisocyanate (TDI) blocked epoxy resins have grown to more prominence and wider acceptance for practice due to their much improved corrosion resistance, adhesion and throwing power over most carboxyl-bearing (polyester, epoxy, etc.) anodic binders. Such blocking may also be additionally accomplished using —OH bearing acrylates. Curing results from chain-extension by thermal polymerization and inter molecular urethanization involving:

(i) the hydroxy acrylic moieties and
(ii) the free isocyanate groups and the acrylate —OH groups or OH groups released on the epoxy resin chain units, respectively.

Due to superior corrosion resistance, higher coulomb yield, better film adhesion and surface levelling, improved edge protection, reduced volatile emission and improved throwing power, more than 95% of all running electro-coating tanks in the world are based on CED technology. Sustained research and development efforts have led to more efficient curing at lower temperatures (130–150°C) with reduced oven emission, better edge coverage through improved rheology control and overall cost advantage.

5.7 Immersion Plating

When the substrate metal is less noble than the plating metal, or can be rendered so by approximate complexing agents in solution, an immersion deposit may be formed as exemplified by the well-known system:

$$Fe + Cu^{2+} \rightarrow Cu + Fe^{2+} \quad (5.1)$$

Most such immersion plating processes are unsuitable practically due to the fact that the deposit formed is powdery and non-adherent.

Tin is deposited on aluminium alloy pistons by immersion in alkaline stannate solutions. The process is commonly used in all automobile manufacturing units. Copper-tin alloys are applied to steel wire (liquor finish) as a drawing lubricant and for colours in such items as paper clips, hair pins, etc. Gold and some other precious metals are also frequently applied by immersion plating techniques.

5.8 Electroless (or Auto Catalytic) Plating

Auto catalytic plating is better known as *electroless plating*. It is defined as the deposition of a metallic coating by a controlled chemical reaction catalyzed by the metal or alloy being deposited. Electroless plating is almost as old as the modern electroplating industry. In electroless plating, the reduction of metal ions to the metal element is achieved by using a chemical reducing agent, while conventional electroplating uses externally supplied electrons as the reducing agent.

Immersion plating is actually an electrochemical replacement process, while *electroless plating* is representative of a *truly chemical reduction process*.

The most widely used among the truly electroless plating processes is nickel plating, in which nickel ions in solution are reduced to the metal by a reductant used in the bath.[4] The deposition process is catalyzed by certain metallic surfaces, including the deposited metal itself; so, once initiated, the process turns autocatalytic.

Electroless plating has several advantages over electroplating:

(i) virtually unlimited throwing power
(ii) little or no excess deposit at high points, and
(iii) the ability to coat the inside surfaces of tanks of, any size which is difficult or impossible to achieve with conventional electrolytic techniques

The principal disadvantage is the relatively high cost. The reducing agent commonly employed, viz sodium hypophosphite, an expensive reagent is consumed in substantial quantities. Electrolytic processes are generally preferred; but electroless methods widen the range of possibilities. Commonly, electroless coating bath formulations contain nickel chloride, sodium hypophosphite and one or more hydroxy acids such as lactic acid or glycolic acid. Temperatures of the range 60–100°C are commonly employed.

Next in practice and importance to electroless nickel is electroless copper, particularly for plating non-metallics (plastics) and in plating for printed circuitry. Electroless copper competes with electroless nickel for plating on plastics. Such platings are applied on specified substrates for achieving an attractive appearance (decorative), protection, special (conducting) surface properties and for enhancing the engineering and mechanical properties of the substrate surfaces. The list of articles plated for attractive appearance is almost never ending: automotive and air craft parts, electrical appliances, plumbing fixtures, office furniture, fire arms, handles and knobs, photographic equipment, luggage/handbag frames, pens and pencils, costume jewellery, office equipment and many others.

After 1950, the growth of electroless coating has been phenomenal and is the direct consequence of:

(i) the discovery of nickel-phosphorous alloys by electroless plating
(ii) the growth of the electronics industry, especially, printed circuits, and
(iii) the large-scale introduction of plastics into everyday life.

5.9 MICROENCAPSULATION

Microencapsulation is the process of coating any finely divided solid or liquid material to form tiny particles of sizes ranging from several tenths of a micron to several thousands of microns. In its simplest form, a *microcapsule* is a small sphere with a uniform wall around it. The material inside is known as the *core, internal phase* or *fill*, while the *wall* is called the *shell, coating* or *membrane*. There may be microcapsules that don't resemble a simple sphere. The core may be a crystal, an adsorbant particle, an emulsion or oil, a dispersion of solids or a suspension of smaller microcapsules.

Depending on the nature of the core and the coating materials, the thickness of the coating and the size of the microcapsules formed, microencapsulation allows:

(a) *The formation* of solid entities from oil
(b) *The protection* of drugs and vitamins from outer environments of moisture, oxygen (air) and light
(c) *The delayed escape* or *slow release* of core substrates, restricting volatilization and loss on evaporation
(d) *The masking* of the disagreeable taste and odour of the core substrate and improving handling of sticky materials, thus making drug administration easy and comfortable, and
(e) *The controlled and sustained release* of fertilizers, pesticides and herbicides, and drugs and pharmaceuticals. This ensures controlled and continuous chemical and drug release or delivery at the desired rates into the outer environment or within the body of a patient possible. It helps achieve a prolonged therapeutic effect after the administration of a sin-

gle measured dose, thus making the endeavour in each case not only much more cost effective but also minimizing or virtually eliminating the adverse effects of overdosage.

The coating (polymeric) materials may be water-soluble, such as gelatin, gum arabic, poly (vinyl pyrrolidone), carboxymethyl cellulose (CMC), methyl cellulose, poly (vinyl alcohol), poly (acrylic acid), polyacrylamide, etc., or water-insoluble, e.g. ethyl cellulose, nylon polyamides, EVA copolymers, and silicones. Natural resins such as shellac or such items as paraffin, bee-wax, stearic acid, stearyl alcohol or glyceryl stearates may also be used for coating powders of clay, $CaCO_3$, etc. that are commonly used as *activated fillers* in polymer compounding. Biodegradable polymers such as microbial poly (3-hydroxy alkanoates), poly (lactic acid), gelatin, starch, etc are useful in the microencapsulation of food and medicines.

5.9.1 Core Wettability

Wetting the core of microcapsules is commonly achieved by a process called *coacervation*, the formation of a separate polymer-rich liquid phase from a polymer solution, e.g. by adding a non-solvent. In coacervation coating, the consequential property is the wettability of the core by the coacervate. As long as solid particles are wet, they are frequently easier to coat than when the core material is a liquid. If a liquid material is highly insoluble in the coacervate-forming solution, proper wetting is likely to be difficult. Sometimes, both solid and liquid cores pose wettability problems.

5.9.2 Core Solubility

In a coacervation system, it is critical that the core not only be insoluble in the solvent for the polymer, but also that the polymer not partition largely into a liquid core. In some cases, cores can have some solubility in the outer polymer solution. In spray-coating to form microcapsules, it is possible to coat a water-soluble, solid core with a wall polymer taken in aqueous solution. This is because the water present evaporates so rapidly under coating conditions soon after the spray droplets hit the cores that there is little penetration or dissolution of the core into the wall-polymer.

The permeability of the microencapsulating polymer and its thickness should permit the release of the core chemical through the crust into the outer environment at the desired rate. The microcapsules must be able to tolerate the rigours of the handling conditions, but at the same time may be required to break or puncture above a threshold pressure (as in the case of some cosmetic items or carbonless copying paper). The nature and thickness of the wall-polymer and the capsule size determine the release rate, elasticity and friability.

The tacky nature of the wall-polymer and the temperature and concentration conditions under which it would so appear are significant controlling features. Spray-coating turns problematic if the wall-forming polymer solution produces strings or filament-like deposits under process conditions, instead of forming film-like deposits. T_m, T_g, the degree of crystallinity and the rate of crystallization, wall-deforming rates, etc. are some of the important controlling features during microencapsulation.

5.9.3 Phase Separation/Coacervation

In some microencapsulation processes, the core material is first suspended in a solution of the wall material. The wall polymer may be induced to separate as a liquid phase (by adding a non-solvent, decreasing the temperature or by additing a phase inducer, i.e. another polymer more soluble in the solvent); incompatibility between the two polymers may cause the first polymer to separate as a polymer-rich liquid phase called a *coacervate* and the process is called coacervation.

As the coacervate forms, it wets the suspended core particles or droplets (if liquid) and finally coalesces into a continuous coating on them. The final step is the setting and isolation of the microcapsules, which is probably the most difficult step in the process. Mixtures of gelatin from different sources, or a mix of gelatin and gum arabic are used at appropriate pH conditions or at a relatively low or near-room temperature such that they together interact to separate as viscous, liquid micro droplet coacervates. If the core particles are easily wet by these micro droplets, a wall of the liquid/swollen coacervate forms on the core particles. The wall so formed may be hardened using minor proportions of formaldehyde or glutaraldehyde leading to cross-linkage. Finally, the liquid is cooled to $\leq 10°C$, pH adjusted to ≥ 9.0 and the microcapsule suspension filtered, washed and dried. The importance of carbonless copying paper had once stimulated microencapsulation studies and practice by many stationery companies.

In carbonless business forms, a dye intermediate, such as crystal violet lactone is microencapsulated to form fine particles ($< 20\ \mu m$ in diameter). Those are deposited as a thin layer on the underside of the top sheet of paper. The receiving sheet or 'copy' is coated with another reactant, such as acidic clay. Such microcapsules resist ordinary handling pressure, but break under high local pressures of pen or pencil tips. When the capsules on the sheets facing each other break, the two chemicals react, producing a dyed copy of the original on the surface of the copying paper at the bottom.

5.9.4 Air Suspension Coating

Such coating of particles by solutions or melts of coating items gives better control and process flexibility. The fine particles are coated while remaining suspended in an upward moving (heated, if necessary) air stream and supported by a perforated plate at the base having different patterns of holes under an inner cylindrical insert at the centre and between the inner insert cylinder and the outer walls of the encompassing cylinder, as schematically shown in Fig. 5.2.

As the particles rise upward by the pressure of the air entered through the bottom perforated plate and distributed in the space above, they encounter a

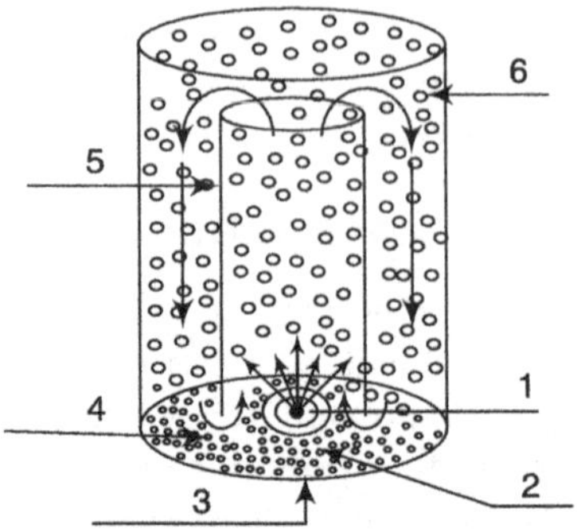

FIGURE 5.2 Principle and practice of air suspension coating: (1) coating spray nozzle, (2) pneumatic nozzles, (3) air distribution, (4) air flow, (5) coating partition, (6) coated particles recycle

fine spray of the coating solution entered into the system from the centrally placed spray nozzle at the bottom of the inner cylinder. Only a small, controlled amount of solution is allowed in. The solvent is driven off and the particles are nearly dry and tack-free as they are blown up; the dried particles fall back onto the outer part of the bottom plate, which may be drawn in again and the cycle is optionally repeated a second or third time to get the desired wall thickness for the microcapsules formed. Particles as big as tablets and as small as 100–150 μm can be easily coated. Since many thin coating-layers are sprayed into all surfaces of the randomly suspended/oriented particles, a uniform coating is attained easily, even on irregularly shaped particles.

5.9.5 Centrifugal Extrusion

A jet of core liquid is surrounded by a sheath of wall forming solution or melt in such a process. As the jet is subsequently ejected into and moves through the air, it breaks into droplets of core, each duly and efficiently coated with the wall solution. While the droplets are in flight, a molten wall may solidify or the solvent for the polymer be evaporated and driven out from the encompassing wall solution. A rotating (jet) head containing concentric nozzles are commonly used. A high production rate (upto 20–25 kg of microcapsules per hour per nozzle) can be produced and achieved. Heads having 15–25 nozzles are used. This process is suited to produce particles of 400–2000 μm diameter. Since the drops are formed by the break-up of a liquid jet, the process is suitable for liquids or slurries.

For microencapsulation of food items, drugs and pharmaceuticals, or veterinary products, only approved wall-forming polymeric materials should be used. Microencapsulation of flavours and essences offers advantages in making candies, specified food items, perfumed goods and surfaces, etc. Spray drying has proved to be an excellent and economical way to form particles containing evenly distributed flavours and scents in a polymer matrix.

5.10 Powder Coating

Much like water-borne coating systems and different types of radiation-cure coating systems, powder coating, developed in the mid- or late-1950s and grown to wide favour, acceptance and practice since the 1970s, is recognized as an environment-friendly and energy-efficient coating process. Powder coatings are currently in popular use as both thermosetting and thermoplastic industrial finishes free from solvents. They are delivered in finely-divided powder form and all techniques of application, primarily on metal surfaces permit reclaim, recovery and recycle of the portion of the powder that misses the target.

A powder used in a powder coating is commonly prepared by mixing and grinding together the base resin or polymer and some additives in the form of pigment fillers and extenders, plasticizers and flow promoters, stabililzers, curatives and catalysts, etc. The common techniques of mixing and blending, some shown in Fig. 5.3, include *melt mixing* (using two-roll open mills, banbury mixers, extruders (single-screw or twin-screw types) and *dry blending* equipments (a vertical drum fitted with a properly designed stirrer, eccentrically rotating drum, or a horizontal drum fitted with high-speed rotating, helically-designed blades, e.g. the Henshel mixer or the "ribbon blender". One may also use a *solution* and/or *a slurry system* followed by *spray drying* the respective systems to fine powder. Melt mixing is usually followed by the grinding or pulverizing of the lumps, cakes or granules into fine powders using a hammer mill, ball mill, attrition grinding and fluid-energy mill or by employing cryo-grinding techniques. Powder coatings of superior appearance cannot be obtained by dry blending because particle sizes of the powders can not be regulated. The pigment-dispersion uniformity also falls short of the desired level; thus, dry blending compositions are much inferior than melt-mixed compositions. Highly brittle or friable melt-mix compositions (e.g. some acrylics, polyesters, phenolics and amino resins) can be ground to fine and desirable sizes in a fluid energy mill. This mill utilizes a high-velocity gas (usually air) stream to make the particles have impact with each other and with the equipment (mill) wall. The gas streams allow rapid cooling of the powder that may get heated up due to the friction and shear involved. The powder obtained is then separated into very fine and bigger grains; for powder coating, the fine grains constituting the major fraction are favoured, and the very fine or bigger grains are separated and reserved for applications employing other technologies.

The application techniques used for powder coating are:

(a) Electrostatic spray
(b) Fluidized bed
(c) Electrostatic fluidized bed, and
(d) Tribo

For applying a powder coating on a metal substrate, the appropriate surface preparation must be done first. The substrate surface is cleaned by removing dust, dirt, oil, grease and any adsorbed gases or chemicals. Phosphate treatment

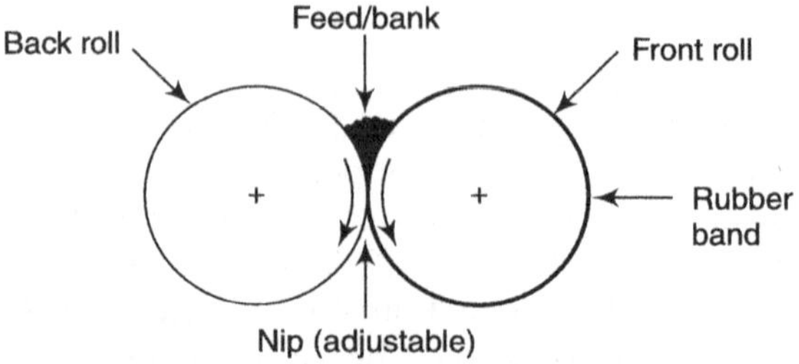
Back roll
Feed/bank
Front roll
Rubber band
Nip (adjustable)

(a)

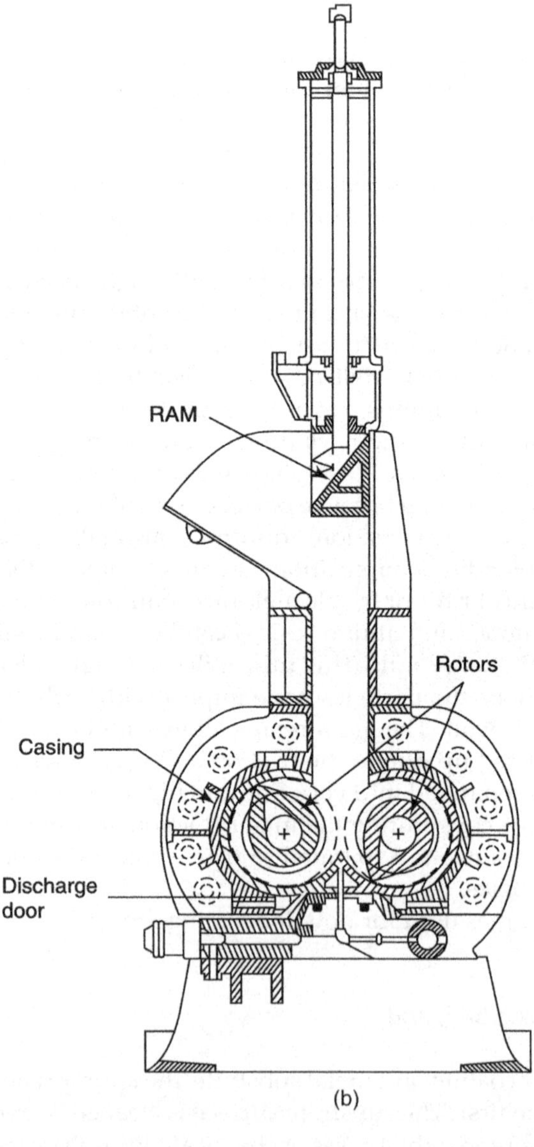
RAM
Rotors
Casing
Discharge door

(b)

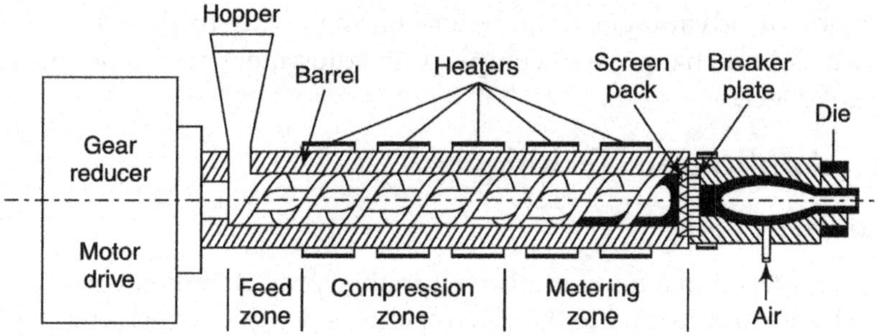

FIGURE 5.3 Some mixing equipments: (a) two-roll open mill, (b) banbury mixer, (c) screw extruder

of the metal surface improves corrosion resistance. For galvanized substrate surfaces, treatment with zinc phosphate or chromate has been proved to be highly helpful. Aluminium substrates are usually given a multistage chromate conversion to achieve maximum protection. In most cases, no primer coat is necessary; the selected powder is applied directly on the pretreated metal surface. The surface preparation is much the same employed for electroplating or electroless plating.

5.10.1 Advantages of Powder Coating

(1) Powder coating is much like a paint, but it is a dry process and does not require any viscosity adjustment or control. Powder coats are used straight from the pack

(2) Being solvent-free, powder paints are eco-friendly and free from the hazards of atmospheric pollution solvent storage and handling. The process does not entail any effluent disposal problems

(3) Liquid paints containing 50%–60% solvents and diluents pollute the environment, especially the application area, due to rapid solvent evaporation that at the same time defies recovery and reuse, creating a huge waste. Powder paints are endowed with high efficiency (coating efficiency upto 98% commonly achievable) and low material loss of <2%–3%. Powder coating is also substantially energy efficient and economical

(4) No primer coat is necessary for powder paints even though a primary coat may be helpful in improving adhesion. A powder paint is in fact a one coat system, offering mechanical properties, edge-protection and chemical or corrosion resistance superior to the prospects offered by common liquid (solvent-based/emulsion) paints

(5) Powder paints are more easily applied, less dependent on skilled operators and free from such problems as dripping, sagging and wrinkling, compared with conventional liquid paints

The major disadvantages of powder coating relate to their being less amenable to colour changes and composition adjustment and sometimes the orange peel effect.

5.10.2 Application Techniques

(a) Electrostatic Spray

The powder, placed in a hopper adjacent to the spray chamber, is picked up by an air stream and carried to the electrostatic spray gun, which charges the powder particles on emission and causes them to deposit on the earthed object. Fusion and curing is then done in an oven. Any powder that escapes deposition on the work piece is collected, reclaimed and reused. A schematic diagram is shown in Fig. 5.4. The advantages of this technique are:

(i) Thin substrates can be conveniently coated
(ii) A film thickness of 30–120 μ can be achieved in one application
(iii) Intricately-shaped articles can be coated
(iv) There is no need to preheat of the work piece

The main disadvantages are that the equipment cost is higher than the fluidized bed method and the coating rate is somewhat low.

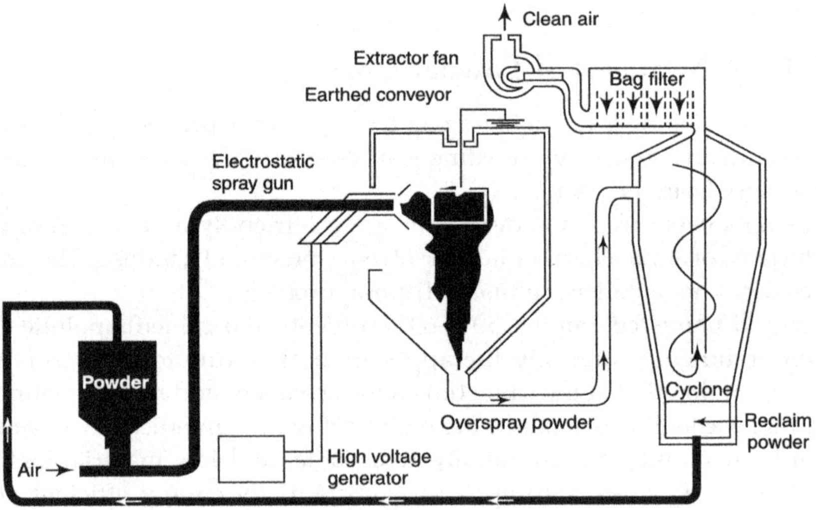

FIGURE 5.4 Powder coating by electrostatic spray technique

(b) Fluidized Bed

This technique is generally confined to cases requiring a relatively thick (200–500 μ) coating in one application, such as for coating valves for fluid (gas/liquid) pipelines where a durable, tough and corrosion/wear-resistant finish is essential.

The fluidized bed essentially consists of a porous membrane or base plate through which a controlled, low-pressure air supply is fed and maintained.

The air thrust serves to fluidize the powder kept within the tank, thus making the system behave much like a liquid and allowing the ready entry of the objects to be coated into the fluidized bed, as shown in Fig. 5.5.

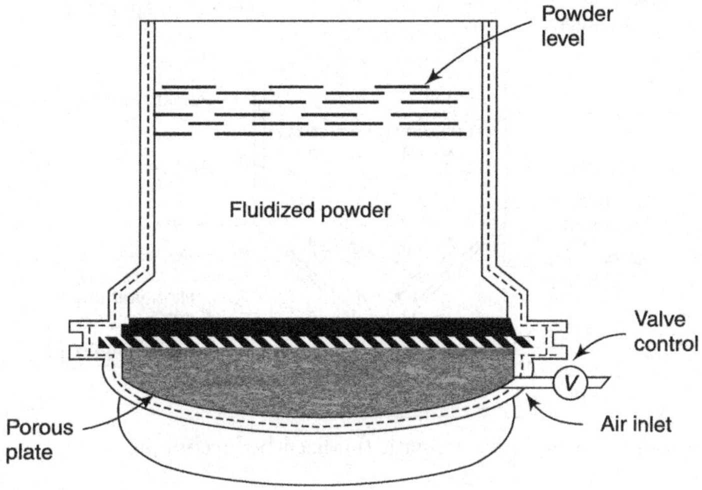

FIGURE 5.5 Powder coating by fluidized-bed technique

Before immersion in the fluidized bed the objects are preheated to a temperature above the fusion point of the powder. The coating thickness is controlled by the time of immersion in the bed and the preheat temperature of the object. In some cases, the preheat temperature may be sufficient to cure the polymer, while for other cases, post-heating may be necessary.

The process is endowed with the advantages that an exceptionally thick coating can be obtained in a single immersion and stoving cycle and the initial plant cost is low. The process is sensitive to the size and the shape of the workpiece. The process is especially useful in coating objects with a high surface-to-mass ratio, and for fabricated wire goods with sharp edges and intersections such as dish washer baskets and refrigerator shelves. The size of the parts that can be coated is obviously limited.

(c) Electrostatic Fluidized Bed

Here, the fluidizing container and the porous plate must be constructed of a dielectric material. In modern models, the fluidizing air is ionized by a high-voltage source before its entry and passage through the porous plate, as illustrated in Fig. 5.6.

Electrostatic fluidized beds are shallow. The parts to be coated are grounded and passed over the bed; the charged powder is attracted to the grounded parts. The parts may be rotated as they pass over the charged cloud of powder to facilitate uniform deposition.

The electrostatic fluidized bed coating is ideal for the continuous coating of webs, wires, fencing and related objects that are fabricated in continuous

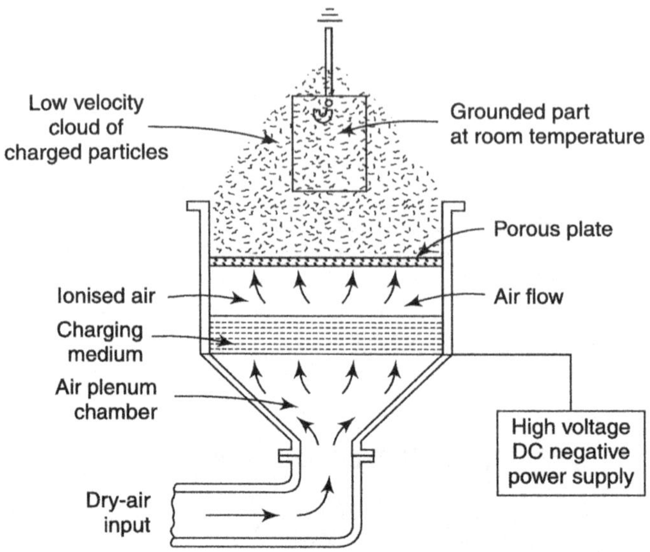

FIGURE 5.6 Powder coating by electrostatic fluidized-bed technique

lengths and are essentially two-dimensional. Sometimes two electrostatic fluid beds are used and the continuous material is passed between them so that both sides are coated simultaneously.

(d) Tribo Technique

When a powder, after being brought into contact with a surface, is caused to move along that surface, the powder particles acquire an electrical charge due to friction. This development *of frictional charge* is a function of, among many parameters, the chemical and physical natures of both the powder and the surface of contacts, the relative humidity and the powder-particle velocity. The powder is fed to the application gun from a fluidized-bed powder hopper, through a variable-choke air injector; no electrical controls are required. The tribo guns as such applicators are known, are advantageous due to their low cost (no high-voltage generator needed), good spread and penetration over the object, and good coating thickness control. However, specially formulated powder is required for coating using a tribo gun and the charging efficiency falls with increase in relative humidity and time of use.

5.10.3 Powder-coating Resins and their Coating Characteristics

The important thermosets used for powder coating include epoxy, acrylic, phenolic, polyurethane, polyester or hybrids (epoxy-acrylic, epoxy-urethane, epoxy-polyester, epoxy-phenolic, etc.). The important thermoplastics in use are nylon polyamides (nylon 11 in particular), linear polyesters (terephthalic acid-1, 4-butane diol based product in particular), copolymers of ethylene-chlorotrifluoro ethylene (E-CTFE), poly (vinylidene fluoride) (PVDF), poly (phenylene sulfide) (PPS), simple polyolefins (PE and PP) or their blends, cellulose esters (cellulose acetate-butyrate in particular) and poly (vinyl chloride) (PVC). Plasticized PVC coatings are commonly used and they are softer and more flexible than any other powder coating materials—thermoplastic or thermosetting. Almost all PVC powder coatings are applied by the fluidized-bed coating process. The resistivity of plasticized PVC powders is low compared to other powder coating materials and the applied powder quality loses its electrostatic charge; PVC powders are therefore, seldom applied by the electrostatic fluidized bed process. Polyolefins have not captured a significant market because of their poor adhesion, while cellulose esters, otherwise advantageous, are not preferred for their high costs. Adhesion can be improved for polyolefins by using a primer or an adhesion-promoting polymer (5%–10% for unpigmented or 15% for pigmented systems) called Hercoprime.

Coating-powders based on thermosetting resins far excel and surpass those based on thermoplastic polymers/resins in usage. For use as a powder coating, a resin must possess low melt-viscosity to afford a smooth, continuous film and good adhesion at the application temperature to the substrate. It must develop high strength, toughness and abrasion resistance on curing or setting and show good resistances to heat, chemicals and weathering. Thermoplastics show a melt viscosity that is of a higher order of magnitude than that exhibited by most thermosetting resins at normal application temperatures. Thermoplastics, however, have one advantage; they don't need curing, even though post-heating may be necessary sometimes to complete fusion and the spreading to uniformity of the powder particles. Thermoplastic resins are favoured for applications in coating wires, fencing and other applications where the process involves continuous coating at high line speeds. Tables 5.2 and 5.3 project the property ranges of some thermoplastic and thermosetting powders used for powder coating and the properties of the relevant coatings.

TABLE 5.2 Some thermoplastics and their property range in powder coating

Property	*Thermoplastic*					
	Vinyls	*Poly-amides*	*Polyes-ter*	*Polyethy-lene*	*Poly-propylene*	*Cellulo-sics*
Primer required	Yes	Yes	No	Yes	Yes	Yes
Particle size (μ)	50–100	< 100	< 105	< 150	< 100	< 105
Preheat temp. °C	290	310	300	230	250	280
Post-heat temp. °C	230	250	250	200	210	220
Specific gravity	1.20–1.35	1.01–1.15	1.30–1.40	0.92–0.94	0.89–0.92	1.15–1.35
Hardness, shore D	30–55	70–80	75–85	40–50	40–60	80–90
Surface appearance[a]	S	S	L-OP	OP	S	L-OP
Gloss 60°	80	70	90	70	80	85
Adhesion [b]	G–E	E	E	G*	G–E*	G–E

(Contd.)

Table 5.2 (*Contd.*)

Resistance [b] to:						
Impact	E	E	G–E	G–E	G	G
Abrasion	P	G–E	G	P	F	G–E
Weathering	G	G	E	P	P	G
Humidity	E	E	G	E	E	F
Solvent	F	E	F	G	E	F
Acid	E	E	G	E	E	F
Alkali	E	E	G	E	E	P

[a]S = Smooth, OP = Orange peel, L-OP = Limited orange peel
[b]G = Good, F = Fair, P = Poor, E = Excellent
*Poor adhesion without primer

TABLE 5.3 Some thermosettings and their property range in powder coating

Property	*Thermosettings*			
	Epoxy	*Polyurethane*	*Polyester*	*Acrylic*
Fusion range °C	120–200	160–210	150–220	120–210
Cure time (min)	1–30	10	10	10
At temperature °C	240–0120	200	200	200
Surface appearance[a]	OP	L-OP	G	G
Adhesion	E	E	E	E
Hardness	H–4H	H–2H	H–2H	H–2H
Gloss 60°	70–90	30–90	50–90	80–95
Flexibility[b]	E	E	E	F
Resistance[b] to:				
Impact	E	G–E	G-E	F
Humidity	G–E	G	G	G–E
Weatherng	P	G–E	E	G–E
Acid	G	F	G	F
Alkali	G	P	F	P
Solvent	G	F	F-G	F

[a]OP = Orange peel, L-OP = Limited orange peel,
[b]G = Good P = Poor, F = Fair, E = Excellent, H → 4H-increasing order of hardness

Powder coatings are viewed as fusion-coating processes because, at some point during the coating process, the powder particles must be fused or melted (and also cured for thermosets). This is commonly accomplished in well designed ovens, even though infrared and resistance or induction-heating methods are also in use. One limitation of powder coating is that the substrate to be coated must be able to withstand the high temperature required for the fusion and curing of the coating polymer/resin. The major powder-coated end products include domestic appliances, auto-components (cars, bicycles, motor cycles), electrical equipments, computer hardware, pumps, bathroom fixtures, fans, false ceilings, furniture, fire extinguishers etc. The usage patterns of powder-coating and relevant resins are shown in Table 5.4.

TABLE 5.4 Usage pattern of powder coating and coating resin powders

Application/area	*(%)*	*Polymer Type*	*(%)*
Furniture	30–35	Epoxy	50–55
Appliances	8–12	Epoxy-polyester	10–12
Building industry	20–22	Polyester	10–15
Automotive	18–22	Vinyl	5–10
Machinery	7–8	Acrylic	6–10
Others	5–10	Others	5–10

5.11 Coated Fabrics

Waterproofing of fabrics meant for clothing or footwear, made by spreading the latex or gum from the rubber tree on the fabric and then drying the web over smoke, was known to the people of the American continents long before the western world woke up to modernity during the Industrial Revolution. Modern textile or fabric coating using rubber (NR) had its beginning with the first patented process of making rain coats by sandwiching a layer of rubber dissolved in coal-tar naphtha between two layers of cloth, as developed by the Scotsman, Macintosh (1923). The process and the product have undergone a sea change and further development after the discovery and development of rubber vulcanization (1839–1843) using sulfur by Charles Goodyear in the USA and Thomas Hancock in the UK.

Since then, a wide range of many other polymers—rubbery and resinous—have been developed and used to make coated fabrics. They are synthetic rubbers or elastomers[5,6] (polychloroprene, SBR, EPDM, NBR, silicone or blends of elastomers, PU lattices and aqueous dispersions), vinyl and acrylic polymers and resins, phenolic and amino (urea/melamine) resins, plasticized PVC and PVC-based plastisols and organosols, copolymers of selected acrylic and methacrylic esters prepared by solution or emulsion polymerization techniques, and resorcinol formaldehyde resin (latex or dry bonding) systems. There are also radiation (electron-beam/UV) curable resin (an unsaturated polyester optionally mixed with a styrene monomer, or urethane acrylate/epoxy acrylate) systems much in demand for use in inks for graphic arts and printing, in lamination, packaging and the electronic industry, but their demand and popularity in textile coating are still low.

Coating fabrics[7,8] from cotton, jute, polyester, rayon, nylon, viscose, etc. (sometimes blended or unionized) using polymers in latex or aqueous dispersion forms is favoured over coating them using polymers in organic solvents because of environmental concerns and pollution control rigours. Still, solution-based coatings are practised for certain systems because of eartain technology advantages. Further, to discourage or do away with solution-based coatings, growing in acceptance and popularity are hot-melt coating processes and those involving plastisols.

5.11.1 Applications of a Primary Adhesive Layer

The adhesion of the polymer to the textile substrate, according to the adhesion mechanisms detailed in Chapter 3, is an important aspect of coating technology, specially when put to dynamic use. The irregularities of the textile substrate surface causing mechanical interlocking of the elastomeric or resinous coatings are fibre ends, twists and turns and crimps of the yarns and interstices of the weave patterns, particularly for cotton fabrics/yarns made from staple fibres. Fabrics made from synthetic fibres, normally produced in continuous filaments are smooth and have comparatively poor adhesion. Hydrogen bonding provides significant cumulative contributions for cellulosic (cotton, viscose and acetate rayon) woollen and polyamide-based yarns/fabrics and coatings based on urea/melamine resins, polyesters and elastomeric polyurethanes. Acetate rayon, nylon and polyester based fabrics/yarns/cords, being made from smooth, continuous filaments, require an adhesive pre-treatment, usually using a resorcinol-formaldehyde latex (RFL) system, a poly (vinyl pyridine) (PVP) latex or an SBR-PVP mixed latex system having 20%–80% PVP content and 80%–20% SBR content in the mixed latex. The solid content of the dip liquid is adjusted to 15%–20% in most cases, leading to an add-on of 5%–15%, depending on the nature of the textile item (fabric or yarn) used; finally, water is removed by heating (110–140°C) the item for a short period (1–2 min) and additionally, heat setting, leading to a higher degree of polycondensation for RFL application to create plenty of reactive (methylol) sites[5]. These reactive sites then offer improved adhesion with the elastomeric/resin (plasticized) coating that is finally laid on the pre-treated textile item.

5.11.2 Dry Bonding

One can also use dry bonding systems[6] where the adhesion-promoting chemicals and additives are directly compounded with the elastomers. The additives in common use are resorcinol (2.5 phr), a formaldehyde-donor such as hexa (hexamethylene tetramine) (1.5 phr) and a bit (upto 10 phr) of hydrated or fumed silica of fine particle size, besides other fillers. Resorcinol and hexa find enough scope and time to react and form the methylol-bearing R F resin that migrates to the interface for bond formation at vulcanization temperature (140–150°C), silica acting as a retarder for the sulfur-accelerator vulcanization process commonly accomplished by using semi-ultra/delayed-action

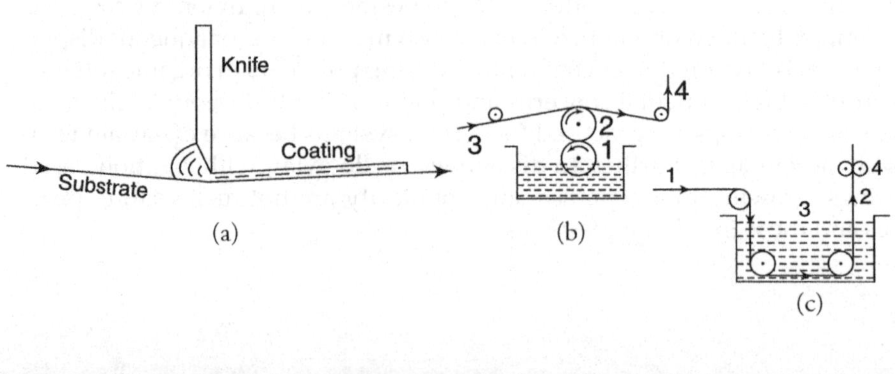

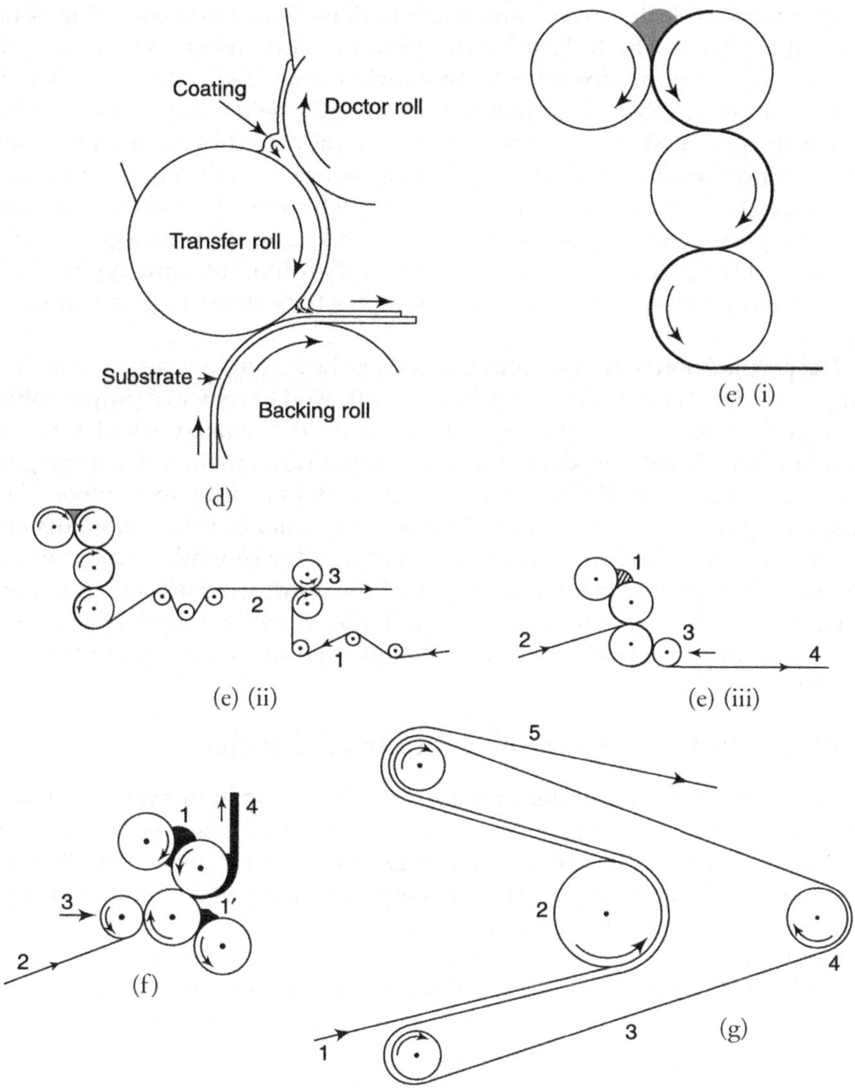

FIGURE 5.7 Layouts of some common techniques for the formation of coated web or for fabric coating: (a) *Coating with floating knife;* (b) *Kiss coating:* (1) pick-up roll, (2) applicator roll, (3) web for coating, (4) coated web; (c) *(Liquid) dip coating:* (1) substrate web, (2) coated web, (3) dipping tank, (4) squeeze rolls; (d) *Reverse roll coating;* (e) *Calendering:* (i) a four-roll (inverted L) calender for sheet-making, (ii) in-line lamination assembly: (1) fabric, (2) calendered sheet and (3) laminating (pressing) rolls; and (iii) inclined calender for making rubberized fabric: (1) rubber bunk, (2) fabric to be coated, (3) pressing (laminating) roll and (4) coated fabric; (f) *Coating on both sides of fabric (Z-calender):* (1) and (1′) rubber bunk, (2) fabric, (3) press-roll, (4) both side coated fabric; and (g) *Rotocure system:* (1) coated fabric for curing, (2) steel drum, (3) steel belt, (4) tension roll (5) coated fabric (cured)

accelerators or selected accelerator combinations[7,8] (sulfenamide, thiazole and guanidine types). The RFL or RF dry bonding systems are widely used for constructing fabric reinforced industrial rubber products such as tyres, beltings (conveyer belts, V-belts and power transmission belts) and hoses[6]. Rubberized (rubber-coated) fabrics are also widely produced by techniques called skim coating (employing contact/low pressure) or frictioning (employing higher positive pressure) on calendar rolls. Such rubberized fabrics are useful in making raincoats as well as baby care and patient-care sheets useful in homes and hospitals. The principles and process lines of some typical fabric coating techniques, including roll assemblies for calendering are shown in Fig. 5.7.

The preparation of rubber dough precedes the actual coating process. Small cut pieces of duly compounded rubber are allowed to soak in a proper solvent or solvent mixture for certain specified hours. The soaked, swollen mass is then put into closed kneaders, where kneading turns them into a homogeneous dough; this may take 10–12 h. Various types of coating techniques,[8,9] viz. (liquid) dip coating, knife coating, kiss coating, roller coating (for emulsions), Zimmer coating, (hot-melt) extrusion coating, dry powder coating and sophisticated calendering are used; some of them are schematically illustrated in Fig 5.7. A calender is used to coat a polymer directly onto the fabric or for making an unsupported film or web that may be subsequently laminated on to a fabric.

5.11.3 Vulcanization of Rubberized Fabrics

Batch curing by hot air under pressure or autoclaving under saturated steam are two options for vulcanizing rubber-coated fabrics. Hot air curing needs more curing time because the heat transfer coefficient of air is lower than that of steam. Further, oxygen in air may come into play as a degradative agent Hence, autoclaving is the preferred process.

Continuous drum cure, also known as *rotocure,* is also used as seen in Fig. 5.7. In this process, the curing of the coated fabric is accomplished by suitably placing the fabric in contact with a rotating steam-heated drum. In fact, a steel band or belt is allowed to run over nearly two-thirds of the circumference of a slowly rotating steam-heated steel drum, the moving steel belt remaining pressed with the coated fabric being run over, in hard pressed conditions over the heated drum. As the pressed rubberized fabric sheet slowly moves with the rotation of the heated drum, vulcanization to optimum degree is achieved.

5.12 Important Applications of Coated Fabircs

5.12.1 Rainwear and Foul Weather Clothing

These are prepared from coated fabrics that block the pores of the textile material by a compact polymeric coating to form a physical barrier to wind and water. They are not breathable as such, and are relatively inexpensive. Generally, a single-textured or double-textured (sandwiched), rubberized fabric is used. PU, PVC or neoprene coated fabrics are in use. Neoprene makes the coated fabric relatively heavy while PVC-coated fabrics have poor low temperature properties. PU-coated nylon is most favoured because of its lightness, the thin coating and excellent low temperature flexibility (for an average area density of 100–250 g/m^2, the PU coating is only 10–30 g/m^2). They are used by the military for cold weather clothing (jackets, trousers, caps, etc.).

5.12.2 Breathable Fabrics

The purpose of these fabrics is to provide comfort as well as to offer protection in foul weather. Such fabrics have demand as sports and leisure wear. Breathable fabrics must possess high water vapour permeability, water proofing character and the ability to withstand high wind pressure; they must be durable, launderable and often heat sealable. They find application as protective clothing for the services. Woven cloth made from 0.05–1 denier polyurethane, polyester, viscose, acrylic or polyamide microfibres are used with a water-repellent finish of silicones or fluoro copolymer emulsion. Microporous PU, polyamide and PTFE coatings on cotton/viscose with a primary adhesive coat, for durability, have been developed.

5.12.3 Synthetic (Artificial) Leather

There was a need to develop synthetic leather due to periodic shortages of leather as well as for economic considerations. Natural leather is avaliable in different sizes and thicknesses; there is no way one can set a standard for natural leather. Synthetic leather, in the form of vinyl/rubber-coated fabrics have been developed and have grown in demand for the footwear, soft luggage and upholstery industries.

Expanded, soft vinyl (PVC)-coated fabrics developed in the 1950s, often termed synthetic leather, provide feel and draping qualities similar to or even better than natural leather. Such expanded vinyl fabrics have a base layer of knitted fabric, an intermediate layer of cellular PVC and a wear-resistant top coat of plasticized PVC. Such coated items are also known as leather cloth, and are usually produced by transfer and calender coating. PU-based fabrics are superior to vinyl-coated ones in leather-like applications considering durability and low temperature flexibility with little chance of embrittlement with time due to plasticizer loss, as in the case of the vinyl coated products. A top-coat of PU is cast on a release paper; this is followed by a coating from a mixture of a chain-extended polyol/polyester system through a reaction with selected diisocyanates and then using a blowing agent in the mix. The composite laminate is foamed and cross-linked at an appropriate high temperature and then finally laminated on a fabric base. The foamed leather so laid on the fabric is then separated by removing the release paper from the top coat.

5.12.4 Inflatables

The low permeability of certain polymeric coatings have rendered relevantly coated fabrics the materials of choice as light-weight, flexible containers for both gases and liquids. The containers for gases are known as *inflatables* and provide buoyancy. The liquid containers are used as storage vessels for water, fuels and the like. As collapsible storage containers, coated fabrics are particularly suitable for transporting liquids by land and sea. Base woven fabrics from high-tenacity nylon yarns (800–1000 d) are commonly used for this purpose.

5.12.5 Tarpaulins

Tarpaulins are used as covers to protect various commodities from weather damage. They are used to cover seeds and food grains, machineries, cement, etc. for storage in the open or during road transportation. Canvas covers, generally made from heavy-duty cotton fabrics with a rot-proof wax-emulsion treatment, once in widespread use, have recently been replaced almost completely by plasticized PVC-coated fabrics called tarpaulins. The advantages of tarpaulin are lightness, strength, waterproofness, resistance to tear, puncture and abrasion, flexibility and durability. Neoprene, PVC-NBR blends, Hypalon or PU-coated tarpaulins are favoured for covers to be used in contact with oils.

5.12.6 Carpet Backing

The fixing of a backing fabric to a carpet imparts certain important properties such as dimensional stability, tuft binding, resistance to water, reduced pilling, etc. Carpet backing may be limited to simple application followed by curing of an adhesive (say SBR latex) layer to the back of the carpet, without bonding secondary fabric; a secondary backing fabric, usually jute or polypropylene, may be bonded to the back of the carpet by an adhesive. One may also provide foam backing, where the back of the carpet is given a thin cushion of foam as its integral part.

The adhesives used may be NR latex, SBR latex, EVA or PVA emulsions in combination with selected additives, including fillers/extenders and viscosity-adjusting thickeners, and a frothing agent. Different methods are used for laminating the backing (jute/PP) fabric. One may apply an undercoat of NR/SBR latex on the back of the carpet and a suitable adhesive (SBR latex/PVAc emulsion) on the jute fabric surface by roller coating; bonding the two by running them between laminating press rolls and allowing the adhesive layers to develop some froth, let them dry, set and cure, as shown in Fig. 5.8.

5.12.7 Flocking

Flocking is the process of applying and vertically fixing short fibres (flocks) on an adhesive-coated substrate. The substrate may be woven or knitted fabric, leather, polymer film or paper. Flocking applied electrostatically yields a

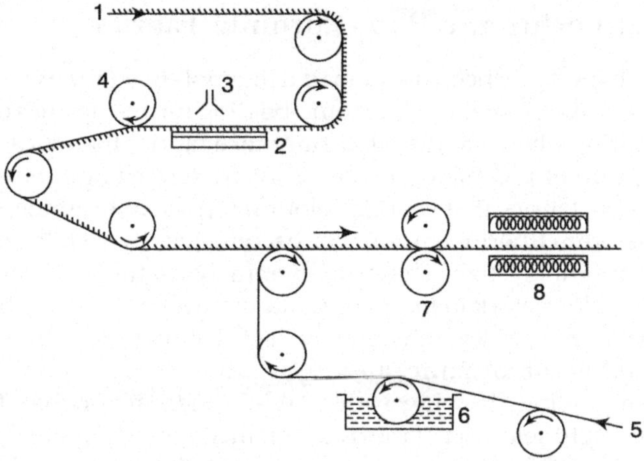

FIGURE 5.8 Layout for carpet backing: (1) carpet, (2) supporting plate, (3) spraying of frothed latex, (4) control roll, (5) backing fabric (jute, PP), (6) adhesive application, (7) press rolls and (8) oven drying

variety of items with aesthetic appeal in the form of velvets, draperies, carpets, artificial furs and suede. The most common short fibres or flocks (0.1–0.6 mm long) are nylons, acrylics and viscose. Flocking provides an easy and economic means of producing of webs of pile texture. The short (flock) fibres are introduced into a high-voltage field; as charge carriers they are transported to the horizontally laid and earthed adhesive-coated fabric surface so as to be anchored on that surface vertically at right angles. The fabric surface bearing the adhesive layer is vibrated by a set of rollers. The short fibres that penetrate the adhesive layer are held in place and retained to form a dense pile. Excess flock fibres escaping anchorage on the adhesive layer are removed by suction. The flocked fabric item is then allowed to pass through a tunnel drier to dry and set the adhesive layer. It is cooled and the excess flock is removed by brushing and applying suction again, as illustrated in Fig. 5.9.

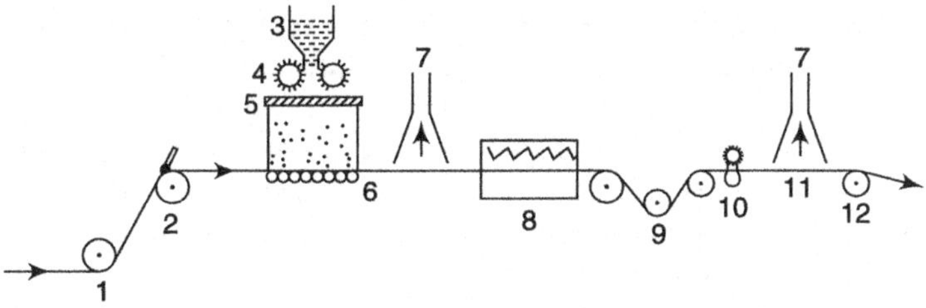

FIGURE 5.9 Layout for flocking: (1) fabric, (2) (knife on roll) adhesive coating, (3) hopper containing short fibres, (4) rotating brush, (5) electrostatic depositor, (6) vibrating rollers, (7) suction, (8) drying oven, (9) cooling rolls, (10) light brushing, (11) suction, and (12) winding

5.12.8 Thermochromic/Photochromic Fabrics

High-tech fibres have lately been developed using polymer (adhesive) binder systems of selected dye pigments or perfume-bearing micro-capsules on them. The perfume bearing fibers are useful as fragrance fabrics that release a perfume on the rupture of the microcapsules. (say, by scratching), the dye pigment fibers provide fabrics that change colour or shade or brightness with a change in temperature (thermochromic) or become readily visible on illumination in photochromic system. These are extremely useful on apparel or on the helmets of people at work in mines or while controlling traffic in the dark; likewise, photochromic paints are variously useful at night on highways.

A good number of organic and inorganic compounds display thermochromism[3], some showing irreversible, others reversible, thermochromism. By the right selection of the dye system, it is now possible to obtain thermochromic effects over temperatures between –30°C and 100°C. Using a thermochromic composition of 1 part crystal violet lactone (dye) and a combination of 4 parts benzoyl-4-hydroxy benzoate with 20 parts stearyl alcohol (able to induce reversible thermochromism), microcapsules may be obtained by coacervation in an aqueous gelatin-gum arabic system. The microcapsules dispersed in PU emulsion are then coated on the fabrics by dip coating and drying. The coated textiles then exhibit reversible thermo-chromism, turning blue at temperatures > 55° C and being colourless at or below room temperature. A binder other than PU may also be used. For a photochromic effect, a photochromic dye or colourant is used instead.

5.13 Flexible Film Former and Flow Promoter-cum-Plasticizer for High-performance Cement Concrete

This application infuses unbelievable and maybe even unsurpassable advantages[10] in the construction industry such as:

(1) Fast mixing to uniformity of cement concrete of standard compositions (stone chips, sand and cement) using a substantially lower proportion of water, thus permitting the saving of 15%–20% water,
(2) Fast release of entrapped air and adsorbed elements (oils, greases, etc.) from the surfaces of solid, sandy, cementitious, stony particles
(3) Prompt wetting of the concrete mass and effecting easy flow and fast spreadability of the concrete admixture
(4) Quick hydration, setting, hardening and more than optimum strength development in 24 h
(5) Development of high compactness, high bulk density, very low porosity and narrow pore size distribution and high cure strength development in 7 days.

Further, much higher reinforcing effects and minimum volume change are observed on the optimum use of an aqueous solution of a water-soluble, high-polymeric material. This solution is prepared by the aqueous copolymerization at 30–40°C of a mixture of several oligomeric (hydroxy) polyether acrylates (overall, 1 mole), acrylic/methacrylic acid (0.75–0.85 mole) and butyl/ethyl/methyl acrylate (0.15–0.25 mole) using appropriate doses of $K_2S_2O_8$ and $Na_2S_2O_4$ as aqueous redox initiators for 24–48 h, while maintaining the pH at 7.0–8.0 and yielding a liquid mass of solid content about 15%–20%. The oligomeric (hydroxy) polyether acrylates are initially prepared by ester interchange or interesterification reactions between assorted, mixed polyethylene glycols or hydroxy-terminated polyether oligomers, using them in nearly equal molar proportions (total 1 mole) and methyl acrylate (1 mole) over a period of 24–48 h employing a transesterification or interesterification catalyst that is very effective at room temperature.

The high performances of the copolymer solution diluted to the overall concentration of only 0.05%–0.2% in a relatively low volume of water for concrete mixing arises as a consequence of its

(i) exceptionally high wetting and efficient film forming, spreading and encapsulating capacity
(ii) prominent surface active (soapy) character
(iii) high tack and adhesion, and
(iv) exceptional ability to form salts between its carboxylate groups and the di- or polyvalent metallic (Ca^{2+}, Mg^{2+}, Fe^{3+}, Al^{3+}) ions present in the solid constituents (cement, sand with a bit of clay as contaminant and stone chips) of the concrete mix.

Moreover, the methacrylic/acrylic acid-(hydroxy) polyether acrylate copolymers help form networks through chemical anchorage on surfaces of the solid components of the concrete mass and on the surface of the reinforcing steel rods, forming ionic bonds, thereby acting not only as process aids, but also as great property improvers for the reinforced concrete systems. In the presence of the plasticizing copolymer, the concrete turns low on brittleness and high on flexural and compressive strengths, impact resistance and resistance to cracking and crack propagation.

The water-soluble hydroxy-terminated polyether acrylate-acrylic acid-acrylate copolymers mentioned may provide excellent heat, light and UV-resistant, curable coatings (using an aldehyde or a diisocyanate as the curative) for protecting RC roof-tops; they may also be used as additives to concrete admixtures for achieving less porous and compact casting, fast hydration and setting, enhanced strength development and durability for the concrete structures.

References

1. Lambourne, R ed, *Paint and Surface Coatings: Theory and Practice,* Ellis Horwood, Chikester, 1987.

2. Schmidt, A X and C A Marlies, *Principles of High Polymer Theory and Practice,* McGraw-Hill, New York, 1948.
3. Kirk-Othmer, *Encyclopedia of Chemical Technology,* 3/E, Vol. 6, 8, John Wiley and Sons, New York, 1981.
4. Pearlstein, F, *Modern Electroplating,* 3/E, WB Harding and F W Lowenheim eds, John Wiley and Sons, New York, 1974, 710–747.
5. Buswell, A G and T J Meyrick, *Rubber Industry,* August, 1975, 146–151.
6. Hofmann, W, *Rubber Technology Hand Book,* Hanser, Munich, 1989.
7. Wake, W C and D B Wooton, *Textile Reinforcement of Elastomers,* Applied Science, London, 1982.
8. Datta, B, *Coated Fabrics in Rubber Manufacturing Technology,* A Bhowmik et al ed, Marcel Dekker, New York, 1994, 473–501.
9. Sen, A K, *Coated Textiles: Principles and Applications,* Technomic, Lancaster, 2001.
10. Ghosh, P, Unpublished Work, Polymer Study Centre, 'Arghya', Kolkata, 2001.

Chapter

6

Tribology and Tribological Coatings

Tribology is the field of science and technology dealing with contacting surfaces in relative motion—meaning thereby that it deals with phenomena related to friction and wear.

In industrialized societies, there is a growing need to reduce or control friction and wear for several reasons, such as to extend the life of machinery and bio-systems, to make engines and devices more efficient, to conserve scarce material resources, to save energy and to improve safety.

Historically, these aims were achieved by design changes, selecting improved bulk materials, or by utilizing lubrication techniques. Bulk material changes often involve applications with ceramics and polymers, while lubrication techniques cover the use of liquid lubricants—such as mineral or synthetic oils, including silicone oils, or solid lubricants—such as graphite powder, stearic acid or other fatty acids or their calcium, zinc or lead salts and molybdenum disulfide.

Recently, tribologists have made increasing use of another approach to friction and wear control: using surface treatments and coatings. This has led to, and to some extent is fuelled by, the growth of a new discipline called *surface engineering*. The growth has been encouraged by two main factors:

(a) The development of new coating and treatment methods that provide coating characteristics and tribochemical properties that were previously unachievable, and

(b) Recognition by engineers and materials scientists that the surface is the most important part in many engineering components. It is at the surface that most failures originate by either wear, fatigue or corrosion. The surface also has many functionally important attributes, not necessarily confined to mechanical or chemical properties but the magnetic, electronic and optical characteristics.

Mechanical components and tools today are facing higher performance demands. The use of surface coatings opens up the possibility of a material design catering to specific properties. The substrate material can be designed for strength and toughness while the coating is made resistant to wear, thermal load and corrosion. Tribologists developed an understanding of the behaviour of surfaces in contact, thus providing a theoretical basis for the prediction of desirable surface attributes even before fully optimized coatings were available.

It is against this background that high-performance coatings and treatment methods are being developed; they are already making a significant impact. Devices and bearing systems that operate under near-vacuum conditions, as in space mechanisms or satellites, engine components operating under hot, corrosive and erosive conditions (as in aero-gas-turbines), would not function without advanced tribological coatings.

6.1 Tribological Coatings

Tribological coatings are those that are thin enough that the substrate material plays a role in determining the friction and wear performance. This definition excludes coatings that are so thick that there is little or no influence of the substrate on the tribological behaviour, and the coating, in effect, acts as a bulk material. Weld deposits are a typical example of such thick coatings, outside the preview of tribological coatings.

The coating processes can be conveniently divided into four generic groups—*gaseous, liquid* (*solution or molten*) and *solid*—depending on the state of the depositing phase. For tribological coatings we shall concentrate on the gaseous state processes that are currently attracting much scientific and commercial interest.

To meet tribological requirements, the coated surface must posses a suitable combination of properties including hardness, elasticity, shear strength, fracture, toughness, thermal expansion and adhesion. The coating surface, the coating material, the coating-substrate interface and the substrate material must have the property attributes shown in Fig. 6.1.

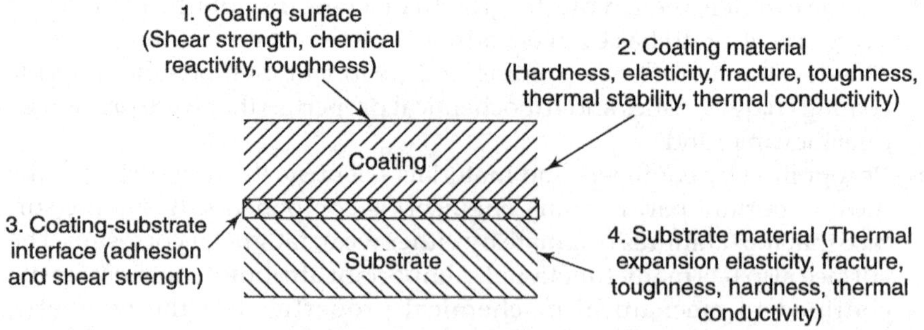

Figure 6.1 Property attributes of different zones of a coating-substrate combination

The properties required by the substrate and the coating involve material strength and thermal attributes determined by their composition, microstructure as well as the porosity and homogeneity of the material. At their interface, the adhesion and shear strength of the junction is important. At the surface of the coating, the chemical reactivity and the roughness must be considered in addition to the shear strength.

A primary problem in surface design is that many desired properties, viz. adhesion at the interface with no surface interaction with the counter-face or high hardness and toughness of the coating cannot be easily obtained simultaneously. Increased hardness and strength are usually concomitant with decreasing toughness and adherence. For this reason, the final coating design is always a compromise between different technical requirements, the economical requirements relating to the properties of the coating system and requirements relating to the deposition of the coating onto the selected end products.

The whole coated surface system with its functional parameters can be considered a composite system to be optimized to gain the maximum benefit. This, in essence, specifies the fundamental principle of surface engineering, which is the design of the surface and the substrate together as a system to give a cost effective performance enhancement of which neither is capable on its own.

It has been established since many years that the tribochemical reactions at the contact interface are critical in many tribological systems. Often, these lead to the formation of stable compounds after a period of contact and rubbing, which then control the ensuing friction and wear behaviour. Advanced surface engineering techniques allow the design and production of such compounds as surface layers, prior to in-service use, thereby ensuring accurate control of the tribological system. It is not only the sliding mechanism that is improved by such possibilities, it may also possibly modify the thermal behaviour, altering both thermal and chemical diffusion effects.-This in turn leads to the possibility of improving the efficiency of machines, with the consequential economic advantages of good tribological design.

An illustrative example is the estimated potential improvement in the efficiency of pumps which, in various forms, are reported to consume roughly 10% of the total energy in industrialized countries and upto 20% in process industries. Improved coatings and coating material selection procedures lead to decreased friction and wear. This enables efficiency to extend the life of machinery and provide considerable reduction in overall energy consumption, apart from reducing the costs of in-service failure or maintenance downtime. It follows that an important objective for tribological coatings be the achievement of extended and predictable life of machines and ideally, non-catastrophic failure modes.

6.2 Tribological Coating Methods

Recent developments in tribological coatings are largely a consequence of new coating methods, providing properties unachievable previously. The properties that one looks for are morphology, composition, structure and adhesion, all combining to give the desired life and durability.

The *plasma* and *ion-based* deposition techniques have generated wide interest. The process developments were delayed due to difficulties with the advanced technologies involved:

(1) High current and high voltage technology
(2) Process control and related electronic technologies
(3) Plasma physics and chemistry
(4) Vacuum technology

With the solution of most of these problems, plasma–based techniques now offer considerable benefits to various sectors of engineering. These methods are used to produce relatively thin surface layers such that the substrate and coating combine to provide a tribological performance that neither can achieve on its own. The important surface engineering techniques for tribological coatings may be generally classified as depicted in Fig. 6.2:

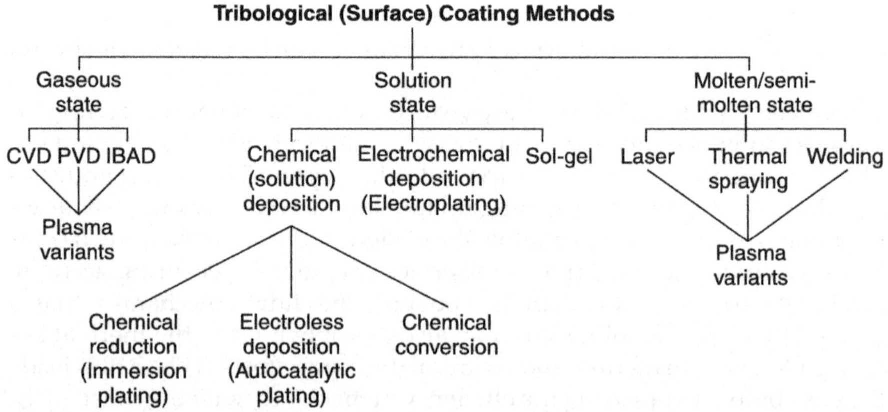

FIGURE 6.2 General classification of surface engineering techniques

Two important characteristic parameters for coating processes are the thickness of the coatings that can be achieved and the deposition temperature. The Abbreviations used for the different process technologies are: PA = Plasma-assisted, CVD = Chemical Vapour Deposition, PVD = Physical Vapour Deposition, IBAD = Ion Beam Assisted Deposition, IAC = Ion-assisted Coating and II = Ion Implantation.

The range of typical coating thicknesses vary from $0.1–10^4$ μm and the temperature range varies from near-room temperature to 1000°C as shown in the charts under (a) and (b), respectively in Fig. 6.3.

6.3 Gaseous State Processes

Gaseous state processes cover surface engineering techniques in which the

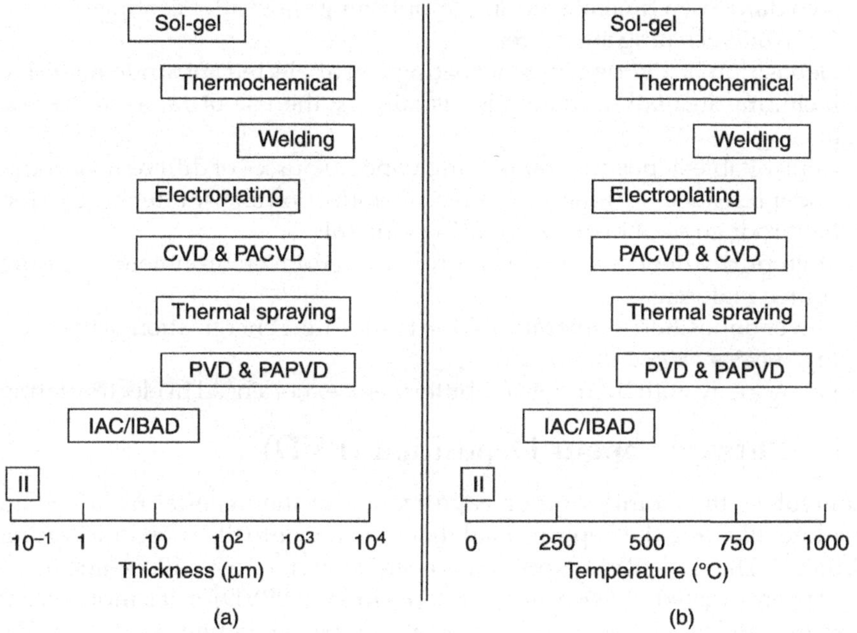

FIGURE 6.3 Typical ranges of (a) coating thicknesses and (b) processing temperatures for surface technologies

coating or surface treatment material passes through a gaseous or vapour phase prior to depositing onto or modifying the surface. The main generic coating sub-groups are Chemical Vapour Deposition (CVD) and Physical Vapour Deposition (PVD). The former utilizes gaseous reagents as the source of coating species, whereas in PVD at least one of the coating species is evaporated or otherwise atomized from the solid state within the coating chamber.

These methods are of considerable interest because they permit the deposition of pure ceramic films. In the case of PVD, this requires a way of increasing the energy of the coating species, most typically by ionizing them and accelerating the ions towards the growing film. This can be achieved by using an ion beam source or by initiating a plasma around the substrate from which ions can be accelerated.

The benefits of plasma-assisted processes, principally relating to PVD but also in several cases pertinent to plasma-assisted CVD, are:

(i) Improved coating adhesion, due to the plasma's ability to clean and preheat substrates by energetic ions and also due to the neutral bombardment of the substrate surface. This mechanism is sometimes termed *sputter cleaning and coating*
(ii) Uniform coating thickness through gas scattering effects and the ability to rotate or displace samples relative to the vapour source during deposition

(iii) Avoidance of a final machining or polishing stage after coating
(iv) Controlled coating structures
(v) Deposition of a wide range of coatings on a selected substrate material, including insulating materials, usually by the use of radio-frequency biasing
(vi) Controllable deposition rates using vapour sources of different varieties under resistance heating, electron beam induction and sputter magnetron
(vii) No production of toxic or polluting effluents
(viii) High purity deposits from a controlled vacuum environment and pure source materials
(ix) Lower deposition temperatures due to the direct energization of the coating species
(x) Doing away with hydrogen embitterment, experienced in electroplating

6.3.1 Physical Vapour Deposition (PVD)

PVD involves the atomization or vaporization of the material from a solid source and allowing the deposition of that material onto the substrate to form a coating.[1,2] The basic PVD process has been known for over 100 years; however, plasma-assisted physical vapour deposition (PAPVD) is of a more recent origin; it started making a major mark as a tribological coating since 1990. The understanding of the process is in the state of being unfolded at a fundamental level; this has led to the modification of the process parameters to provide major property benefits, viz. improved-to-excellent adhesion, structure control and low–temperature ceramic deposition.

The advantages of PAPVD include the possibility of depositing alloy compounds, multilayer compositions and structures, and the ability to vary coating characteristics throughout the deposited film, giving the concept of a functionally graded coating.[3-5]

Resistance heating is employed for evaporating low melting point materials among metals such as aluminum, copper, silver and lead for deposition on a specified surface.

Electron beam guns are becoming increasingly popular, since there is no limit on the melting point of materials that can be evaporated. Among the different types of guns in use, the work-accelerated electron beam gun system and the hollow cathode discharge gun system are widely used.

Induction heating has the advantage that the heat is directed to the melt material rather than to the boat or crucible. This can induce an evaporant-crucible reaction and is particularly attractive for materials such as titanium and aluminium.

Arc evaporation was particularly used for vaporizing carbon, simply by striking an arc between two carbon electrodes. The arc technique is also viewed with interest for evaporating metals such as titanium. Ionization of only the depositing material is a prime advantage; it also provides some design freedom in the layout of the arc source system. A possible disadvantage is the deposition of macro-particles in the coatings.

The sputtering technique leads to the production of atomistic coating species through the bombardment of a source or target by ions and accelerated neutrals, usually of argon. A major advantage of this approach is that alloys can be sputter-deposited with retention of their composition. The technique commonly incorporates magnetic confinement to enhance and control the extent of ion and neutral bombardment of the target and therefore enhance the deposition rate.

6.3.2 Ion Implantation and Ion-beam-assisted Deposition

Ion surface interaction effects include the possibility that the ions will have sufficient energy to become embedded within the surface. This can provide a means of enhancing the tribological properties, although not producing a coating. Typically, this is achieved by generating a high energy ion beam in a separate source as shown in Fig. 6.4.

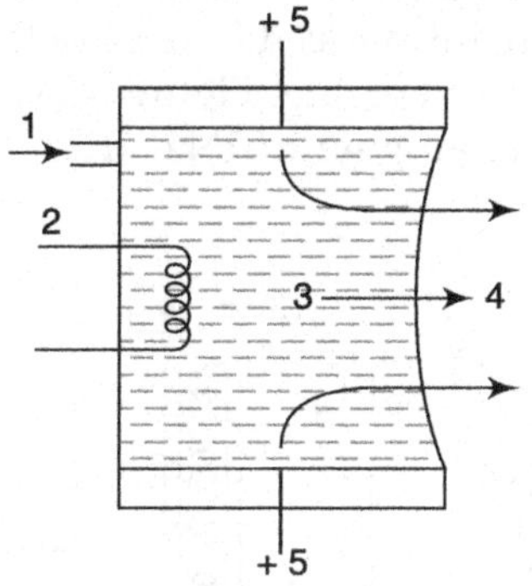

FIGURE 6.4 High-energy ion beam generation: (1) gas inlet, (2) filament, (3) plasma, (4) ion beam, (5) anode

The introduction of selected elemental species such as nitrogen, titanium, carbon or yttrium into the surface of many metals, ceramics, cermets or even polymers has been shown to improve resistance to wear, friction, fatigue and corrosion.

In ferrous alloys, mobile interstitial additives such as nitrogen can influence dislocations and thus impede dislocation movement. Also, in certain alloy steels and in other metals such as titanium, it is possible to form stable compound layers or phases such as metal nitrides, which can harden the surface or enhance corrosion resistance by forming say, a protective oxide layer. The implantation of additional species such as yttrium and rare earths has been shown to further control corrosion resistance.

Recent developments to infuse cost-performance benefits involve scaling up the ion beam system or moving to a plasma source for ion implantation or plasma immersion ion implantation, in tune with ion-based thermo chemical processing systems like plasma nitriding or plasma carburizing. There is significant potential that they can be combined to provide two-stage or duplex processes for pre-or post-treatment during plasma-assisted PVD or CVD processing.

Ion-beam-assisted deposition (IBAD), in which deposition and ion irradiation are simultaneously combined, is useful in forming compound film coatings. Both structural features and chemical properties can be modified. The technique is also known as ion-assisted coating (IAC). If the bombarding ions are reactive species, the process is known as reactive ion beam assisted deposition (RIBAD). Sometimes a combination of techniques may be used. For example, a very thin layer may first be deposited by sputtering or electron beam evaporation and then intermixed with the substrate using an ion beam. Further depositions are then combined with ion implantations to produce a final coating over a highly intermixed interface. This type of ion assisted coating, usually carried out with reactive ion beams such as nitrogen, carbon or oxygen, has been termed Reactive Ion Assisted Coating (RIAC).

6.3.3 Chemical Vapour Deposition (CVD)

In the basic CVD process, gases containing volatile compounds of the element or elements to be deposited are introduced into a reaction chamber and condensed on the substrate to form a coating. Figure 6.5 shows a hot-wall CVD layout typically used for tool coating with TiN or TiC.

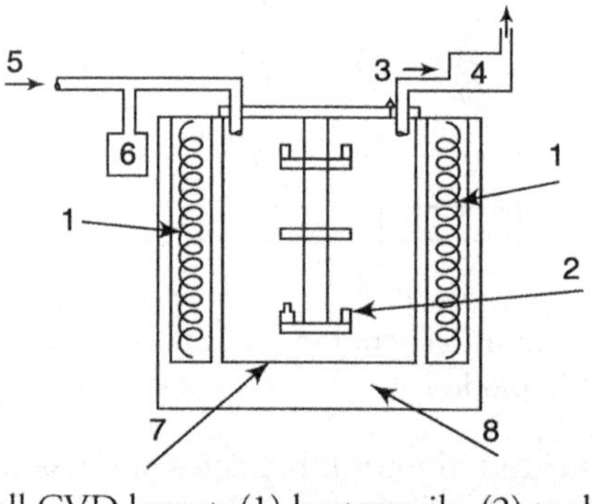

FIGURE 6.5 A hot-wall CVD layout: (1) heater coils, (2) tools, (3) exhaust, (4) exhaust scrubber, (5) carrier gas, (6) $TiCl_4$, (7) stainless steel retort, (8) electric furnace

The deposition pressure in CVD can range from atmospheric to ≤ 1 Pa. There are also various means of assisting the process such as through laser or electron beams or by the ion bombardment of the growing films. There remains a considerable scope for wider use of conventional (non-enhanced), thermally-activated CVD coating methods for tribological applications. The importance of the technique lies in its ability to produce well-adhered, uniform and dense surface layers. The grain orientation and size, the coating composition and its properties can all be varied by selecting appropriate process parameters. The technique can be used to deposit a wide number of wear-resistant coatings such as borides, carbides, nitrides, oxides, carbo-nitrides and oxy-nitrides of almost all the transition metals.

The microstructure of the coating critically depends on the deposition temperature, which for TiN is typically 950°C, but for CVD in general, the temperature may vary from room temperature up to 2500°C. At higher temperatures, various solid-state reactions such as phase transformation, precipitation, recrystallization and grain growth may occur in Zones 3 to 5, as shown in Fig. 6.6.

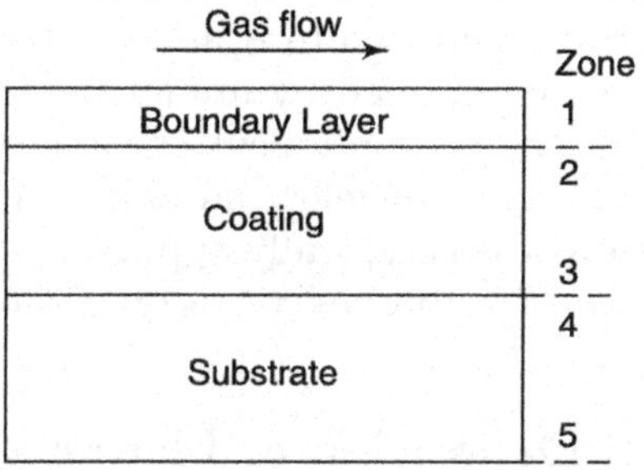

FIGURE **6.6** Reaction zones in thermally-activated CVD

In Zone 4, interdiffusion between the coating and the substrate can lead to the formation of intermediate phases. Such effects are critical to the effective adhesion of the coating. Modifications of the process may allow lower deposition temperatures, control of deposition rate or may permit selective deposition. Plasma-assisted CVD may offer benefits similar to those observed[6–8] for plasma assisted PVD.

6.4 Coating from Solution State Process

The main coating techniques in this category are *electroplating* and *electroless plating* (also see Chapter 5). The solutions are usually aqueous and coating deposits can be produced on metallic or non-metallic substrates. Methods may be categorized as chemical or electrochemical, though this may not be straight forward, as some apparently chemical reactions may in fact be electrochemical.

The chemical reduction process, also known as the autocatalytic process, is exemplified by spraying solutions of metal ions and reducing agents on to a substrate surface, or dipping the latter into the solutions such that the metal coating builds *in situ*. The substrate may be metal or plastics, but the adhesion levels can be poor. The common coating depositions done in this way are silver, gold and copper (also see Section 5.8).

Electroless deposition of nickel, normally with phosphorous or boron additives, is becoming an important coating technology. The additions are derived from reducing agents, viz. sodium hypophosphite (P) or aminoboranes, typically with nickel sulfate. A significant advantage of the electroless nickel process is that it can be used in conjunction with finely divided particles, in the Ni-P bath for example, to produce composite coatings. Examples of suspended particulate materials used to reinforce coatings in this way are SiC,

WC, CrC, Al_2O_3 or diamond.[6,7] Alternatively, solid lubricant particles like graphite, PTFE or calcium fluoride may be added.

Chemical conversion coating methods include phosphating and chromating. The main use of the former is to enhance the corrosion barrier properties of steels or aluminum, prior to painting. The layer, produced by reaction with phosphoric acid, can be either an amorphous or a crystalline phosphate. Sometimes this layer is applied as an aid to the lubrication of moving parts by providing a degree of protection against scuffing and to facilitate running-in. Chromate conversion coatings are applied by immersing in (or spraying a surface with) an aqueous solution of chromic acid, chromic salts, phosphoric acid or other mineral acids. The surface develops an oxide film, which is seated by a metallic chromate. The process is typically applied to steel, aluminum, magnesium, cadmium and zinc.[9] It improves corrosion resistance and can enhance paint adhesion.

6.4.1 Electrochemical Deposition or Plating

Such deposition, also termed electroplating, involves the deposition of a metallic coating on an electrode by a process of electrolysis, whereby chemical changes are produced by the passage of a current. A typical electroplating layout is shown in Fig 6.7. The laws of electrolysis first put forward by Michael Faraday in 1833 still form the basis of the technology; they are:

(i) The amount of chemical change produced is proportional to the quantity of electricity passed

(ii) The amounts of substances liberated by a given quantity of electricity are proportional to their equivalent weight

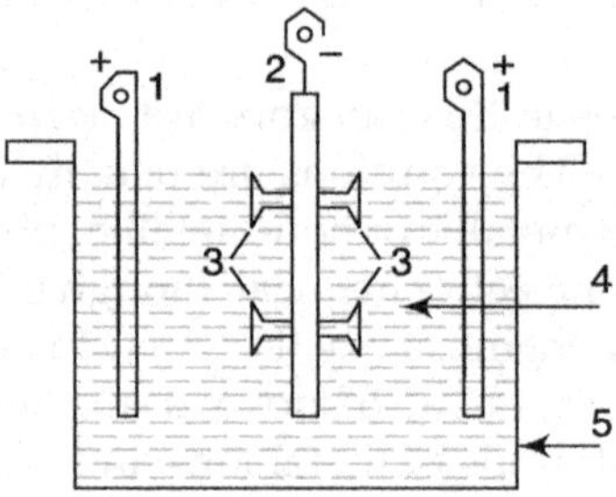

FIGURE 6.7 Typical electroplating layout: (1) anode, (2) cathode rod, (3) surfaces to be plated, (4) plating solution, (5) electroplating bath

The most electronegative metal commonly plated is zinc which, when deposited on steel, provides galvanic protection as does cadmium, though cadmium being toxic, is on the decline in usage. Various alloys—brass (Cu-Zn), bronze (Cu-Sn), and others such as CO-Ni, Fe-Ni, Sn-Ni, Sn-Zn, Sn-Pb—are used as alternatives in the electrochemical coating process (also see chapter 5).

One of the most common electroplated tribological coatings is chromium plate, which is often actually Cu-Ni-Cr or Ni-Cr, the Cr being only a very thin over-layer to prevent the tarnishing of the Ni layer. "Hard chrome" is a term often used for thicker chromium deposits, which have good abrasion and lubricant retention properties.

A few electroplating processes use non-aqueous solutions and some metals have been deposited from solvents such as liquid ammonia. The two main alternatives to aqueous baths are organic solvents and fused salts, though both present problems[12] in terms of toxicity hazards, disposal and cost.

6.4.2 Sol–gel Processing

A sol is a colloidal dispersion of fine particles in a liquid, usually an aqueous system, but sometimes it may as well be an organosol. Sol–gel processing may be carried out by dipping, spraying or spinning and is an emergent coating technology. It involves applying the sol to a substrate, whereupon it undergoes aggregation, gelation and finally dries to form a gel. Typically, the techniques are used for the production of oxide ceramic films, where, after application, the gel is fired or heated above 150°C to leave the ceramic coat. The process has considerable potential not only as a means of producing adherend films of controlled porosity but also as a means of producing spheres of controlled size, which can then be worked with in plasma spraying processes.

References

1. Pauleau, Y, "Physical Vapour Deposition Techniques I: Evaporation and Sputtering" *Advanced Techniques in Surface Engineering*, W Gissler and H A Jehn eds, Kluwer Academic Publishers, Dordrecht, The Nederlands, 1992, 135–179.
2. Moll, E, "Physical Vapour Deposition Techniques II: Ion Plating, Arc Deposition and Ion Beam Deposition" *Advanced Techniques in Surface Engineering*, W Gissler and H A Jehn eds, Kluwer Academic Publisher, Dordrecht, The Nederlands, 1992, 181–197.
3. Mathews, A, *Developments in Ionization Assisted Processes*, J. Vac. Sci. Tech., Vol. A3, 1985, 2354–2363.
4. Rickerby, D S and A Mathews eds, "Advanced Surface Coatings" *A Handbook of Surface Engineering*, Blackie, Glasgow, UK, 1991, 364.
5. Rickerby, D S and A Mathews eds, "Ceramic Coatings by Physical Vapour Deposition" *Review on Powder Metallurgy and Physical Ceramics*, Vol. 4, 1991, 155–195.
6. Agostinov, D, R Favia, P Fracassi and P Lamendola, "Plasma—Enhanced Chemical Vapour Depositon", *Advanced Techniques in Surface Engineering*, W Gissler and H A Jehn eds, Kluwer Academic, Dordrecht, The Nederlands, 1992, 105–133.
7. Bachmann, P K, G Gastner and H Lydtin, "Plasma Assisted Chemical Vapour Deposited Processes" *MRS Bulletin*, Vol. 13, 1988, 52–59.
8. Holmberg, K and A Mathews, "Coatings Tribology—Properties Techniques and Applications, "D Dowson ed, *Tribology Series*, Vol. 28, Elsevier, 1994.

9. Parker, K, "Hardness and Wear Resistance Tests of Electroless Nickel Deposits" *Plating*, Vol. 61, 1974, 834–841.
10. Celis, J P, J R Roos and M de Bonte, *Developments and Applications of Electroless Coatings in Surface Modification Technologies*, T S Sudarshan and D C Bhat eds, The Metallurgical Soc, Penna, USA, 1988, 215–235.
11. Ebdon, P R, Composite Coatings With Lubricating Properties, *Trans. Inst. Met. Fin.*, Vol. 65, 1987, 80–82.
12. Ross, R B, *Hand Book of Metal Treatment and Testing*, Chapman and Hall, London, 1988, 548.

Chapter 7

Miscellaneous Topics

7.1 Classified Adherends and Adhesives

A description of classified adherends and applicable adhesives is given in Table 7.1.

7.2 Factors Favouring Adhesives and Coatings Applications

An idea about performance-cost and other economic advantages of adhesives and coatings applications is given in Table 7.2.

7.3 Testing and Evaluation of Adhesives and Coatings

Studies of a specified material science and technology remain incomplete without the knowledge of the factors and reasons for failure in performance and of approaches and procedures for testing and evaluating the surface properties and bulk properties of the materials used in a given assembly and of the consequential properties of the assembly as a whole.

A comprehensive treatise on approaches for chemical and physical testing, covering procedures for identification and for chemical, mechanical, thermal and spectroscopic evaluations of most structural materials, particularly of rubbers, resins, plastics, fibres, textiles and of polymer blends and composites are available in the author's book on polymer science and technology[1]. The reader would do well to go through the relevant pages of that book for the testing and evaluation of adhesives and coatings as well.

TABLE 7.1 Adherend–adhesive systems at a glance (General Guide for an Adhesive-Joint Designer)

Adherends commonly bonded using adhesives	*Adhesive types for jointing different adherends*
1. **Low-polarity adherends:** (a) Hydrocarbon polymers (rubbers/plastics): Butyl rubber, Hevea (natural) rubber, SBR, polystyrene, PBR, polyethylene and other polyolefins and olefin-copolymers, etc. (b) Halogenated rubbers/plastics: Chlorinated/chlorosulfonated polyethylene, polychloroprene, fluoroplastics/fluoro-rubbers, chlorinated rubber, poly (vinyl chloride), vinyl chloride-vinylidene, Chloride copolymers (Saran), etc.	1. **Wood adhesives:** These are most important in volume and varied applications such as. (a) Phenolic resins, amino (urea/melamine) resins as primary glues or bonding agents for making plywood/paper laminates (b) Animal or vegetable protein glues; once widely used, they are on the decline now with the advancements in modern synthetic adhesive resins/polymers
2. **Medium-polarity adherends:** Poly (phenylene sulphide), Thiokol (polysulphide) rubbers, acrylic/methacrylic plastics, acrylate rubbers, acetal resins/plastics, vinyl chloride-vinyl acetate copolymers, silicones (elastomeric/resinous), etc.	2. **Paper and paper-board adhesives:** (a) The more common adhesives in use in this area are based on starch, dextrin, molasses, agar/gum arabic, etc., all being water-based; animal/vegetable protein based aqueous glues are also in use to make paper cartons, for book binding, packaging, posters, stickers and stamps. (b) Other important adhesives used in this area are poly (vinyl acetate) (aqueous emulsion) and poly (vinyl alcohol) (aqueous), phenolic resin adhesives (alcohol and water based), amino resin adhesives (water-based) or powder form, hydrolyzed shellac (water based), sodium silicate adhesives (aqueous), bitumen-asphalt, PU-adhesives (one-pack/two-pack systems) and PS, EVA, ionomer based hot-melt adhesives, suited for packaging, lamination, book-binding, footwear/luggage industry, etc.

(*Contd.*)

Table 7.1 (*Contd.*)

3. **High Polarity adherends:** Wood: plywood laminates, wood veneers, wood planks/dusts, etc. Cellulose: cotton and related, fabrics, fibres, yarns and cords paper/paper board Cellulosics: cellulose nitrate, cellulose acetate, cellulose acetate-butyrate, ethyl cellulose, etc. (protein, rayon, polyamide, polyester) fibres/yarns/cords; nylons, amino plastics, polyesters and polycarbonates, leather, etc.	3. **Glass/ceramic adhesives:** Silicate adhesives, acrylate/cyanoacrylate adhesives, epoxy/phenolic adhesives, silicones, poly (vinyl butyrals), special rubber adhesives, PU adhesives, etc.
4. **Other adherends:** Rocks, sand, marble, quartz, glass, ceramics, minerals, metals and alloys	4. **Leather, rubber, plastic adhesives:** (solvent-based or emulsion-form) diene rubber and acrylate rubber adhesives, epoxy/PU/phenolic adhesives, heat-sealing or solvent-aided sealing/bonding for thermoplastics, silicone emulsion or silicone rubber adhesives and sealants 5. **Metal and alloy adhesives:** (a) Welding (employing metal welding rods) and soldering for metal-metal bonding (b) Adhesive bonding using epoxy, epoxy-phenolic, epoxy-acrylate and epoxy-urethane (mixed with elastomers) to provide selectively chosen hybrid adhesives, poly (vinyl acetate), poly (vinyl acetals/butyrals) and elastomer adhesives including thiokols

TABLE 7.2 Advantageous features of adhesives and coatings

Adhesives	*Coatings*
1. **Cost factor** would generally favour adhesive bonding wherever it is viewed as a formidable material alternative for jointing two substrates, like or unlike	1. **Surface coatings** represent a very big segment of material science and technology and the chemical industry
2. **Adhesive bonding** permits (a) lighter construction (b) the bonding of an entire assembly in one step with uniform stress-distribution (c) easy application by casting, dipping, brushing, knife-spreading or spraying (d) elimination of drilling, rivetting, welding, soldering and surface finishing steps Adhesive bonding provides environmental protection, a smooth surface contour and vibration resistance, when compared with mechanical fastening	2. **Metallic corrosion** (iron and steel, primarily) amounts to a loss and the cost involvement from all sources comes to the tune of 10^{10} dollars per year. It is a huge loss in terms of materials, energy or human efforts, being also linked with development of safety hazards. Surface coatings protect, decorate and help prevent corrosion and associated losses
3. **Adhesives** may sometimes prove to be more expensive, have limited thermal stability, require rigorous substrate-surface preparation and the aid of jigs, presses and other fixtures to hold parts in place until the bonding adhesive cures or solidifies. However, for paper, thin films, fibres, yarns, cords and non-woven/woven/knitted textiles, foams, glass/ceramics or other brittle materials, adhesive application is the only sensible route to assembly and in such cases, mechanical fastening is often impractical. Adhesive application may prove more advantageous than other methods for jointing dissimilar materials due to unequal thermal expansion rate and failures consequent to that	3. **Powder coatings** are gaining ground over high-solids water/solvent based coatings, due to better surface coverage, high coating efficiency and pollution-free application, needing no waiting for drying; they are ultimately a case of melt-coating

(*Contd.*)

Table 7.2 (*Contd.*)

4. Adhesive consumption across the world since 1960 has been on the rise, increasing 2–4 fold in consumption every two decades. Adhesives have become an integral part of the industrial scene today; only a rare manufactured product today does not make use of adhesives either in its construction or in its packaging, shipping and sales	**4. Advancement** in the scope for material selection, availability and performance profile have enabled modern coatings last considerably longer in service than those prepared some decades ago and repainting is needed less frequently. Additionally, there is enhanced scope today for lowering film thickness with comparable coverage consequent to design of improved parts improved coating formulations and for economy-infusing coating material rheology infusing major upcoming points of developing economic advantages
5. Natural adhesives (bio-adhesives), viz. starch, dextrin, molasses, agar, gum arabic-based and animal/vegetable protein based aqueous adhesives constitute about 30%–50 % of the total use of adhesives. The consumption of natural, rubber-based adhesives (in latex and solution forms) is also significant. The use of synthetic adhesives is in the order, phenolic resin type and hot-melt type ≥ vinyl polymer, synthetic elastomer type > acrylate types; amino resin types > epoxy, polyolefin/olefin copolymer types and others including ionomers, polyamides, polyimides and polyesters. Natural adhesive formulations (bio-adhesives) necessarily contain a low, appropriate dose of a food preservative or biocide to protect them against microorganism attack	**5. Coating systems** are selected according to the requirements of the substrate, keeping factors such as overall cost, process cost, hazards, application flexibility, durability, freedom from pollution and pollution control imperatives in mind
6. Electrically conducting adhesives, whenever required, are made using polymer compounds (resin/rubber-based) bearing 5%–15% freshly made, microfine/nanofine metallic powder (Cu, Al, Steel, etc.) or 10%–30% conducting carbon black (ultra fine or microfine) as fillers. Blending an intrinsically conducting polymer (e.g. polypyrrole, polyaniline, etc.) in an	**6. Coating failures** may occur due to a number of reasons such as: (a) *Cracking,* developed due to shrinkage, thermal shock/fatigue, an oxidative degradation or the aging of the coating (b) *Chalking,* the separating of a loose, removable powder from beneath the coating

(*Contd.*)

Table 7.2 (*Contd.*)

insulating polymer matrix may also provide useful conducting adhesives in melt form or as dispersions using selected solvent systems	(c) *Checking,* the development of fine breaks or micro-cracks in the film, not penetrating to the underlying surface (d) *Erosion* caused by the wear of the finish to expose the substrate or the undercoat (e) *Flaking or scaling,* the detachment of portions of the film coat itself from the substrate surface or from the previously applied paint or undercoat (f) *Blistering,* caused by loosening or lifting portions of coating due to penetration of moisture through the underlying porous substrate (wood, paper, foamed/expanded substrates) or evolution of gas/moisture in localized corrosion attack (g) *Peeling,* the lifting of large film segments from the substrate surface, owing to a moist environment and a mismatch in thermal-induced dimension changes (total expansion/rate of expansion) (h) *Colour fading* due mostly to oxidative, reductive or other chemical changes in the coating, leading to the obscuring of the initial chemical groups, or the appearance of new chromophoric groups, or due to dirt/dust pick up (i) *Loss of gloss,* mostly caused by erosion, dirt/dust and environmental stress cracking (j) *Environmental effects* like oxidation, hydrolytic, thermal and microorganism attack, photodegradation (mainly due to UV-attack), chemical attack (attack by nitrogen oxides, sulfur dioxide, etc.), use of appropriate doses of selected antioxidants/antiozonants (phenolic/aromatic amine

(*Contd.*)

Table 7.2 (*Contd.*)

	compounds and UV-stabilizers (aromatic ketones) are effective in controlling/eliminating oxidative degradation and UV-degradation
7. Adhesive bond design of a plastic and related joint must ensure there is no large-scale chemical reaction between the adherend and the adhesive and within the bond, except adhesive curing; the common bond design criteria are: (a) Uniform stress distribution (b) Self alignment (c) Self fixturing (d) Easy clampability (e) Scope for vertical insertion (f) Scope for horizontal application, and (g) Bond line gap control	
8. No thumb rule exists to predict the mode of potential failures of adhesive bonds; bond failures or durability can be anticipated keeping the combined effects of: (a) adherend-adhesive materials (b) temperature (c) stress level (d) environmental conditions (e) thermal expansion rate mismatch between (unlike) adherends and thermal weakening of the adhesive layer (f) selective migration of base materials or additives (g) bonding process induced degradation of the assembly (h) bond-line geometry, including bond-line thickness and (i) thixotropic character of the adhesive system during application	

7.3.1 Testing of Adhesive and Coating Compositions

For adhesive applications, the base materials and the bonded assemblies must be tested as the coating-substrate surface combine for coating applications. The first group of tests is enormous and diverse because hundreds of individual components may be used in formulating adhesives and coatings. The enumeration of specific tests for each of the major and minor components, viz. monomers, oligomers or polymers, tackifying resins, plasticizers or flow promoters, accelerators, retarders, activators and curing agents, catalysts, fillers, pigments, surfactants/foaming agents, solvents and diluents, gelling agents, thickening agents, etc. does not fall within the scope of this book. The properties of viscosity, storage life and pot life, moisture absorption, aging/weathering resistance, hardness and abrasion resistance, flex endurance, thermal endurance or fatigue, tack, peel strength and cure rates are variously enumerated and documented[1-8] in many text books, handbooks, encyclopediae and appropriate standard test methods such as ASTM.

7.3.2 Adhesive Bond Testing

After an adherend-adhesive bond has been designed, the bond effected and the prototype manufactured, there may be a number of ways for testing bond integrity. The testing may be destructive or non-destructive, short-term or long-term. The tests may be performed on parts or the whole assembly specimen. These test techniques are used to build up useful results and a database to establish quality assurance programmes.

7.3.3 Shear Testing of Adhesive Joints

Stress may be applied to an adhesive joint in shear, tension or a combination thereof. The stress may be applied uniformly or incrementally, where the failure is commonly known as peel and the bond strength measured is the peel strength. The adhesive (or other plastic) bonding methods generally exhibit very high load-at-failure values in shear; so, the simple overlap shear test (ASTM) D 1002 is recognized as the most widely adopted short-term destructive test for adhesive bonds or other plastic bonds.

The adhesive bond strength or adhesion strength can be determined by either a dynamic or a dead-weight method. In the dynamic method, the separated (flexible but inextensible adherend) plies of the specimen are clamped vertically or horizontally to the jaws of the autographic strength-testing machine with a constant rate of traverse; the work of detachment up to a preset length of separation is provided directly by the moving loading device.

In the dead-weight test device with two grips, the separated ends of the adherend plies of the specimen are attached to the two grips, and the deadweight is gradually placed on the lower grip until the two plies (or of a coating on a flexible adherend ply such as a fabric) are separated within a specified rate (say 5 mm in 5 min). The load applied is then recorded to get a measure of the bond strength. The bonds of coated fabrics are likewise evaluated by similar dynamic or dead-weight test methods.

7.3.4 Tests for Coatings

Evaluation of coatings are based on tests for

(i) Colour—Colour measurement or matching is accomplished, depending on visual matching, spectrophotometric or colorimetric measurements or matching

(ii) Hiding power—This is measured in terms of contrast ratio which, according to ASTM, is the ratio of the reflectance of a dry paint film over a black substrate of ≤ 5% reflectance to the reflectance of the test paint equivalently applied and dried over a substrate of 80% reflectance

(iii) Abrasion and hardness—Abrasion (mass loss or mass retention %) testing is done using a special abrading (sand/emery) paper (warped on a rotating cylinder) on which the test coating of a specified size is pressed and rubbed in a programmed manner for a specified period or over a specified length (i.e. using a fixed mode of rotation and period of rubbing); for hardness testing, a standard metal indenter is used under a specified dead-load on the coating/substrate surface and the hardness is expressed in Rockwell, Shore A or Shore D scales

(iv) Adhesion—To assess this property, painted metal or related panels are bent double over mandrels of specified dimension and examined for development of cracks or peel-off to judge loss of adhesion or adhesion failure

(v) Gloss—Gloss refers to specular reflectance or light intensity reflected at the same angle as the angle of incidence. It is measured using appropriate photometers and compared against a reference standard, viewed as a theoretically perfect mirror to which a value of 1000 is assigned. Gloss figures are expressed in parts per thousand and not as a percentage. The practical working standard is a polished black glass plate/tile with a value of 95. High gloss paints would have values of 85–90 measured at an angle of 60°

(vi) Viscosity—A variety of methods are used to measure paint or glue viscosity. One variety, known as *Efflux* or *Flow index viscometer*, measures the time in seconds for a fixed volume of liquid paint/glue to flow under gravity down a specified channel—a higher flow-time means a more viscous paint/glue. The measured time for a given paint is viewed as its *flow index*. Paddle viscometers measure the time needed for a specified paddle to complete a given number of revolutions under a fixed driving force. Aqueous adhesive solutions are commonly formulated so as to give a Brookfield viscosity of 2–5 Pa with a solid content of 30%–50%.

(vii) Exposure—The effects of exposing painted panels to outdoor conditions are a valuable indicator of their performance under service conditions. Painted test panels are exposed vertically or at 45° facing south. At the site of exposure, panels are better prepared to conform to uniformity in application conditions and test conditions. Exposure of panels at different locations (tropical, subtropical, equatorial etc.) and in different seasons are practiced and periodic tests are performed for recording and preparing a comprehensive data-base.

(viii) Accelerated weathering (aging)—Exposure tests, as in the previous point, are slow and time consuming, so accelerated or artificial weathering tests are often more convenient. The test panels are exposed to specified carbon/xenon arc sources or mercury vapour lamps, preferably with selected filters to screen out penetrating short wave radiations not encountered in direct/diffuse sunlight

(ix) Other tests—They relate to the use of
  (a) specified humidity chambers or saturated water vapour chambers for evaluating the short- and long-term effects of moisture
  (b) immersion tests using separate solutions of salts, acids and alkalis of graded concentrations and also simply water at different temperature conditions for assessing hydrothermal degradation under the specified environments
  (c) tests for environmental stress cracking resistance, and
  (d) mildew/microorganism (attack) resistance or rot resistance tests, employing selected culture media or adopting standard soil burial tests

7.4 Metallic Corrosion

An important objective of applying surface coatings (lacquers, varnishes, paints, platings, etc.) on metal surfaces, particularly iron and steel, or other panels and structures is to protect the relevant surfaces from weather attack and prevent weakening due to mass loss as a result of corrosion. A brief discussion of the mechanism of corrosion is worthwhile in this context.

Corrosion is the attack on metal by a sequence of chemical/electrochemical reactions involving the surrounding medium or the environment. Most metals are subject to corrosion, but the large-scale, massive production of iron and steel and the lack of protective capacities of iron oxides—make the protection of iron against corrosion especially important.

Iron is prone to oxidative attack; but under normal weather conditions, the corrosive attack on iron is primarily electrochemical. Known or adventitious metallic impurities present in commercial iron and steel structures initiate the corrosive process. In the presence of moisture and conductive electrolytes, electromotive cells are created between iron and the other metallic impurity sites, which are mostly cathodic in respect of iron. Iron, coming in contact with moisture and excess oxygen, forms ferrous hydroxide. The metal with more positive standard electrode potential becomes the cathode. The initial reaction may be written as:

$$Fe \rightarrow Fe^{2+} + 2e^- \tag{7.1}$$

The released electrons move through the conducting metal substrate to the cathodic (impurity metal) site. If the moist environment at the cathodic surface is acidic, it acts as the driving force for the following electrochemical reaction:

$$2e^- + 2H^+ \rightarrow 2H \rightarrow H_2 \tag{7.2}$$

Under acidic, aerated situations, the cathodic reaction primarily taking place is:

$$2e^- + 2H^+ + ½\,O_2 \rightarrow H_2O \quad (7.3)$$

However, under alkaline situations, the predominant cathodic reaction is:

$$2e^- + H_2O + ½\,O_2 \rightarrow 2OH^- \quad (7.4)$$

The initial (Fe^{2+}) ferrous ion formed and the anion OH^- generated subsequently at the cathode migrate to each other through the medium, forming at some point of time the insoluble ferrous hydroxide:

$$Fe^{2+} + 2OH^- \rightarrow Fe\,(OH)_2 \quad (7.5)$$

Ferrous oxide initially formed oxidizes under the moist condition, utilizing atmospheric oxygen (air) to form rust, which is the brown, hydrated ferric oxide:

$$2\,Fe\,(OH)_2 + ½\,O_2 \rightarrow Fe_2O_3\,.\,H_2O + H_2O \quad (7.6)$$

Further hydration of the ferric oxide may lead to the formation of $Fe_2O_3\,.\,nH_2O$ which is a constituent of rust. For limited oxygen supply, a greenish yellow/brown product considered a mixture of the two iron oxides [ferroso ferric oxide, FeO/Fe_2O_3, i.e. (Fe_3O_4), hydrated forms] remain mixed in the ensuing corrosion product called rust.

Both oxygen and moisture are needed for corrosion to occur. In the absence of liquid water and at a relative humidity < 60%, rusting or corrosion of iron seldom takes place. Under dry conditions, even uncoated steel/iron exposed outdoors may remain unaffected and unchanged for years. Above the critical relative humidity of 60%, condensation occurs in the micropores of even the most highly polished metal surfaces, thus providing the liquid medium needed and enabling the electrolytic process to occur.

7.5 Wear and Tear

The deformation and degradation of materials solely due to mechanical loading (tension, compression, shear, folding, flexing, sonication, vibration, twisting, abrasion) cause wear and tear in them. Exposure to such mechanical actions of low or high intensities, for short or long periods causes minor or major losses in weight, change in dimensions and shine of a given substrate. Such wear and tear may take place under long-time static loading or short-time/long-time dynamic loading conditions, leading to cracking or tearing, loss in sharpness or shine or development of bluntness, fragmentation and fatigue failures. Wear due to the application of mechanical forces is also known as *erosion*.

References

1. Ghosh, P, *Polymer Science and Technology—Plastics, Rubbers, Blends and Composites*, 2/E, Tata McGraw-Hill, New Delhi, 2002.

2. Koshi, H and C L Schneberger, *Adhesives in Manufacturing,* G L Schneberger ed, Mercel Dekker, New York, 1983, Ch. 22, 551.
3. *Test Methods for Pressure-sensitive Tapes,* 7/E, Pressure Sensitive Tape Council, Glenview, III.
4. *Specifications for Adhesives, A Desk-top Bank,* International Plastics Selector, Inc., San Diego, Calif., 1979.
5. ASTM D 1002.
6. Paul, S, *Surface Coating Science and Technology,* John Wiley, New York, 1996.
7. Pocius, A V, *Adhesion and Adhesives Technology,* Hanser, Munich, 2002.
8. Tractor, A A and D Satas, *Coating Technology Handbook,* Mercel Dekker, 2000.

Selected Readings

1. Kinloch, A J, *Adhesion and Adhesive Science and Technology*, Springer, 1987.
2. Marrion, A R, *The Chemistry and Physics of Coatings*, Royal Society of Chemistry, 2005.
3. Paul, S, *Surface Coating Science and Technology*, 2/e, John Wiley, 1996.
4. Petrie, E M, *Handbook of Adhesives and Sealants*, McGraw-Hill, 1999.
5. Pocius, A V, *Adhesion and Adhesives Technology*, Hanser, 2002.
6. Tractor, A A and D Satas, ed, *Coating Technology Handbook*, Mercel Dekker, 2000.

Index

Author's Profile

Premamoy Ghosh, Former Professor and Head, Department of Polymer Science and Technology, Calcutta University, is the author/coauthor of several books and nearly 300 research papers and review articles published in different journals, particularly in the area of polymer science and technology. His research interests cover polymer synthesis and characterization; polymer modification, polymerization kinetics and mechanistics, photopolymerization, polymer end groups and microstructure; graft and block copolymerization; rubber compounding and vulcanization; modification of cotton, wool, jute and related fibres; fibre reinforced polymer (FRP) composites, polymer rheology and polymer morphology; paper and textile finishing; polymer blends and interpenetrating polymer networks (IPN) and conducting polymers. He also contributed chapters in several edited volumes on polymer science and applied chemistry and in the *Encyclopedia of Polymeric Materials*.

Prof. Ghosh is the head of "Polymer Study Centre," instituted by him in Kolkata in 1996. He is a member of the editorial board of the *Journal of Polymer Materials*, M D Publications, New Delhi, and chairman of the editorial board of the journal *Indian Science Cruiser*, published by the Institute of Science, Education and Culture (ISEC), Kolkata. He is a former member of the editorial board of the *Indian Journal of Technology*, CSIR, New Delhi. Prof. Ghosh is a Fellow of the Institute of Materials (London), formerly the Plastics and Rubber Institute (London), and of the Institution of Engineers (India); he is a life member of the Indian Institute of Chemical Engineers, Kolkata and the Indian Association for the Cultivation of Science, Kolkata.

Prof. Ghosh's contributions in Bengali literature include a collection of poems titled "SUPARNA" (2005), a book on polymer science *Polymer Vijnan O Manab Samaj* (2005) and a 350-page memoir "Esho Chirasundar" (2001), depicting his experiences on aspects of education and training, science and society and our hopes and aspirations in the light of the ongoing globalization and the socio political and economic transformations in India and abroad over the past 70–100 years.

Made in United States
Orlando, FL
04 November 2026

90200977R00105